Protected Areas

Conservation Science and Practice Series

Published in association with the Zoological Society of London

Wiley-Blackwell and the Zoological Society of London are proud to present our *Conservation Science and Practice* series. Each book in the series reviews a key issue in conservation today from a multi-disciplinary viewpoint.

Each book proposal will be assessed by independent academic referees, as well as our Series Editorial Panel. Members of the Panel include:

Previously published

Elephants and Savanna Woodland Ecosystems: A Study from Chobe National Park, Botswana
Edited by Christina Skarpe, Johan T. du Toit and Stein R. Moe
ISBN: 978-0-470-67176-4 Hardcover; May 2014

Biodiversity Monitoring and Conservation: Bridging the Gap between Global Commitment and Local Action
Edited by Ben Collen, Nathalie Pettorelli, Jonathan E.M. Baillie and Sarah M. Durant
ISBN: 978-1-4443-3291-9 Hardcover;
ISBN: 978-1-4443-3292-6 Paperback; April 2013

Biodiversity Conservation and Poverty Alleviation: Exploring the Evidence for a Link
Edited by Dilys Roe, Joanna Elliott, Chris Sandbrook and Matt Walpole
ISBN: 978-0-470-67478-9 Paperback;
ISBN: 978-0-470-67479-6 Hardcover; December 2012

Applied Population and Community Ecology: The Case of Feral Pigs in Australia
Edited by Jim Hone
ISBN: 978-0-470-65864-2 Hardcover; July 2012

Tropical Forest Conservation and Industry Partnership: An Experience from the Congo Basin
Edited by Connie J. Clark and John R. Poulsen
ISBN: 978-0-4706-7373-7 Hardcover; March 2012

Reintroduction Biology: Integrating Science and Management
Edited by John G. Ewen, Doug. P. Armstrong, Kevin A. Parker and Philip J. Seddon
ISBN: 978-1-4051-8674-2 Paperback;
ISBN: 978-1-4443-6156-8 Hardcover; January 2012

Trade-offs in Conservation: Deciding What to Save
Edited by Nigel Leader-Williams, William M. Adams and Robert J. Smith
ISBN: 978-1-4051-9383-2 Paperback;
ISBN: 978-1-4051-9384-9 Hardcover; September 2010

Urban Biodiversity and Design
Edited by Norbert Müller, Peter Werner and John G. Kelcey
ISBN: 978-1-4443-3267-4 Paperback;
ISBN: 978-1-4443-3266-7 Hardcover; April 2010

Wild Rangelands: Conserving Wildlife While Maintaining Livestock in Semi-Arid Ecosystems
Edited by Johan T. du Toit, Richard Kock and James C. Deutsch
Edited by Glyn Davies and David Brown
ISBN: 978-1-4051-6779-6 Paperback; December 2007

Reintroduction of Top-Order Predators
Edited by Matt W. Hayward and Michael J. Somers
ISBN: 978-1-4051-7680-4 Paperback; ISBN: 978-1-4051-9273-6 Hardcover; April 2009

Recreational Hunting, Conservation and Rural Livelihoods: Science and Practice
Edited by Barney Dickson, Jonathan Hutton and Bill Adams
ISBN: 978-1-4051-6785-7 Paperback;
ISBN: 978-1-4051-9142-5 Hardcover; March 2009

Participatory Research in Conservation and Rural Livelihoods: Doing Science Together
Edited by Louise Fortmann
ISBN: 978-1-4051-7679-8 Paperback; October 2008

Bushmeat and Livelihoods: Wildlife Management and Poverty Reduction
Edited by Glyn Davies and David Brown
ISBN: 978-1-4051-6779-6 Paperback; December 2007

Managing and Designing Landscapes for Conservation: Moving from Perspectives to Principles
Edited by David Lindenmayer and Richard Hobbs
ISBN: 978-1-4051-5914-2 Paperback; December 2007

Protected Areas: Are They Safeguarding Biodiversity?

Edited by

Lucas N. Joppa, Jonathan E. M. Baillie and John G. Robinson

Registered Office
John Wiley & Sons, Ltd, The Atrium, Southern Gate, Chichester, West Sussex, PO19 8SQ, UK

Editorial Offices
9600 Garsington Road, Oxford, OX4 2DQ, UK
The Atrium, Southern Gate, Chichester, West Sussex, PO19 8SQ, UK
111 River Street, Hoboken, NJ 07030-5774, USA

For details of our global editorial offices, for customer services and for information about how to apply for permission to reuse the copyright material in this book please see our website at www.wiley.com/wiley-blackwell.

Library of Congress Cataloging-in-Publication Data

Names: Joppa, Lucas, editor. | Baillie, Jonathan, editor. | Robinson, John G., editor.
Title: Protected areas : are they safeguarding biodiversity? / edited by Lucas Joppa, Jonathan Baillie, and John Robinson.
Description: Chichester, West Sussex : John Wiley & Sons, Inc., 2016. | Includes bibliographical references and index.
Identifiers: LCCN 2015036766| ISBN 9781118338162 (cloth) | ISBN 9781118338155 (pbk.)
Subjects: LCSH: Protected areas. | Natural resources conservation areas. | Biodiversity conservation. | Wildlife conservation.
Classification: LCC S944.5.P78 P754 2016 | DDC 333.72–dc23 LC record available at http://lccn.loc.gov/2015036766

A catalogue record for this book is available from the British Library.

Wiley also publishes its books in a variety of electronic formats. Some content that appears in print may not be available in electronic books.

Cover image: Getty/Kandfoto

Set in 10/12pt Minion by SPi Global, Pondicherry, India
Printed and bound in Malaysia by Vivar Printing Sdn Bhd

1 2016

Contents

Contributors

J. E. M. Baillie Zoological Society of London, London, UK
M. Barnes Centre of Excellence for Environmental Decisions, The University of Queensland, St. Lucia, Queensland, Australia
School of Geography Planning and Environmental Management, The University of Queensland, St. Lucia, Queensland, Australia
B. Bertzky United Nations Environment Programme World Conservation Monitoring Centre, Cambridge, UK
Institute for Environment and Sustainability (IES), European Commission, Joint Research Centre (JRC), Ispra, Italy
L. Boitani Department of Biology and Biotechnologies, Università di Roma, Rome, Italy
T. M. Brooks NatureServe, Arlington, VA, USA; IUCN, Gland, Switzerland
N. D. Burgess United Nations Environment Programme World Conservation Monitoring Centre, Cambridge, UK
Center for Macroecology, Evolution, and Climate, Natural History Museum of Denmark, University of Copenhagen, Copenhagen, Denmark
S. H. M. Butchart BirdLife International, Cambridge, UK
L. R. Carrasco Department of Biological Sciences, National University of Singapore, Singapore
B. Collen Centre for Biodiversity & Environment Research, University College London, London, UK
C. Corrigan United Nations Environment Programme World Conservation Monitoring Centre, Cambridge, UK
I. D. Craigie ARC Centre of Excellence for Coral Reef Studies, James Cook University, Townsville, Queensland, Australia
C. Davis College of Forestry and Conservation, University of Montana, Missoula, MT, USA
G. Dubois European Commission – Joint Research Centre, Brussels, Belgium
N. Dudley Equilibrium Research, Bristol, UK
R. A. Fuller School of Biological Sciences, University of Queensland, St. Lucia, Queensland, Australia
L. Gurney European Commission – Joint Research Centre, Brussels, Belgium
J. Haas USDA Forest Service, Rocky Mountain Research Station, Fort Collins, CO, USA
A. Hansen Ecology Department, Montana State University, Bozeman, MT, USA
S. Hedges Wildlife Conservation Society, Bronx, NY, USA
M. Hockings School of Geography Planning and Environmental Management, The University of Queensland, St. Lucia, Queensland, Australia

L. Joppa Microsoft Research, Cambridge, UK
D. Juffe-Bignoli United Nations Environment Programme World Conservation Monitoring Centre, Cambridge, UK
S. Kenney United Nations Environment Programme World Conservation Monitoring Centre, Cambridge, UK
N. Kingston United Nations Environment Programme World Conservation Monitoring Centre, Cambridge, UK
L. Krueger The Nature Conservancy, Arlington, VA, USA
J. Loh WWF International, Gland, Switzerland
School of Anthropology and Conservation, University of Kent, Canterbury, UK
K. MacKinnon IUCN World Commission on Protected Areas (WCPA) Cambridge, UK
B. MacSharry United Nations Environment Programme World Conservation Monitoring Centre, Cambridge, UK
L. Maiorano Department of Biology and Biotechnologies, Università di Roma, Rome, Italy
L. McRae Institute of Zoology, Zoological Society of London, London, UK
A. Milam United Nations Environment Programme World Conservation Monitoring Centre, Cambridge, UK
EcoLogic, LLC, Birmingham, AL, USA
E. J. Milner-Gulland Department of Life Sciences, Imperial College London, Berkshire, UK
Department of Zoology, Oxford University, Oxford, UK
M. A. K. Mwangi BirdLife International, Cambridge, UK
H. Nagendra School of Development, Azim Premji University, Bangalore, India
E. Nicholson School of Botany, University of Melbourne, Melbourne, Victoria, Australia
Deakin University, Geelong, Australia. School of Life and Environmental Sciences, Centre for Integrative Ecology (Burwood Campus), Australia
T. G. O'Brien Wildlife Conservation Society, Bronx, NY, USA
Mpala Research Centre, Nanyuki, Kenya
N. Pettorelli Institute of Zoology, Zoological Society of London, London, UK
N. Piekielek University Libraries, 208L Paterno Library, The Pennsylvania State University, University Park, PA, United States
M. Rao Wildlife Conservation Society, Bronx, NY, USA
J. G. Robinson Wildlife Conservation Society, Bronx, NY, USA
C. Rondinini Department of Biology and Biotechnologies, Università di Roma, Rome, Italy
L. Santini Department of Biology and Biotechnologies, Università di Roma, Rome, Italy
J. P. W. Scharlemann United Nations Environment Programme World Conservation Monitoring Centre, Cambridge, UK
School of Life Sciences, University of Sussex, Brighton, UK

D. B. Segan Wildlife Conservation Society, Bronx, NY, USA
G. Shahabuddin Centre for Ecology, Development and Research, Uttarakhand, India
E. J. Stokes Wildlife Conservation Society, Bronx, NY, USA
S. Stolton Equilibrium Research, Bristol, UK
J. Walston Wildlife Conservation Society, Bronx, NY, USA
J. E. M. Watson Wildlife Conservation Society, Bronx, NY, USA
School of Biological Sciences, University of Queensland, St. Lucia, Queensland, Australia
School of Geography, Planning and Environmental Management, University of Queensland, St. Lucia, Queensland, Australia
M. Wegmann Remote Sensing and Biodiversity Research, University of Würzburg, Würzburg, Germany
German Aerospace Centre, Cologne, Germany
S. Whitmee Centre for Biodiversity & Environment Research, University College London, London, UK
Institute of Zoology, Zoological Society of London, London, UK
International Union for Conservation of Nature, Cambridge, UK

Introduction: Do Protected Areas Safeguard Biodiversity?

J. E. M. Baillie[1], *L. Joppa*[2] *and J. G. Robinson*[3]

[1]Zoological Society of London, London, UK
[2]Microsoft Research, Redmond, WA, USA
[3]Wildlife Conservation Society, Bronx, NY, USA

In 1959, the UN Economic and Social Council called for a list of the world's national parks and equivalent reserves to recognise their economic, social and scientific importance and for their role in environmental well-being. The protected area network at the time covered roughly 2,000,000 km^2 and is now rapidly approaching 20,000,000 km^2 (WDPA, 2014). This tenfold increase in protected area coverage over 50 years has been one of the greatest successes in conservation. But protected areas are not an end in themselves, and to a large extent, biodiversity loss has continued unabated. Populations of many species have continued to decline, some species have gone extinct, and the integrity of ecosystems has increasingly been threatened. The world population has more than doubled, as has the human consumption of water, food and energy. Ever-increasing land conversion, carbon emissions, spread of invasive alien species, nitrogen pollution and over-exploitation have driven biodiversity loss. Ecosystems such as forests, coral reefs, mangroves and seagrass beds are on the decline (Butchart et al., 2010; Emmott, 2013), taking with them the species they contain (WWF, 2012). And these trends are not changing; the United Nations projected there will be 9.6 billion people on the planet by 2050 (United Nations, 2013) and the consumption of food, water and energy will more than double, resulting in the conversion of many of the last remaining wild spaces. As human population increases and land is converted or degraded, protected areas will therefore play an increasingly important role in conserving biodiversity. The imperatives of providing resources to meet the needs of an expanding human population while protecting other life forms will surely collide, and there will be growing pressures to develop or exploit

---

Protected Areas: Are They Safeguarding Biodiversity?, First Edition.
Edited by Lucas N. Joppa, Jonathan E. M. Baillie and John G. Robinson.

these remaining protected landscapes. We are now at a point where we must now decide how much space we are going to leave for other forms of life.

Although protected areas are recognised as one of the most effective tools for conserving both terrestrial and marine biodiversity (Bruner et al., 2001; Selig & Bruno, 2010), we have only a preliminary understanding of the extent to which the portfolio of protected areas safeguards biodiversity on a global scale. This book aims to define what is currently known about the current coverage of the global protected area network, the extent to which this network is truly safeguarding biodiversity and steps needed to improve the effectiveness of our protected area portfolio. It is hoped that this will provide a solid foundation for protected area planning in the 21st century.

Strengthening the foundation for protected area planning is particularly important as the world's governments have recently made a commitment to expand global protected area coverage to at least 17% of terrestrial and inland water habitats and 10% of marine areas by 2020, 'especially areas of particular importance for biodiversity…' (Target 11) (CBD, 2010). This coverage is mandated to be 'ecologically representative'. However, before expanding the current global network of protected areas, it is important to first understand how much of the world is currently under protection, which components of biodiversity are within this network and where, how and whether the system is functioning effectively.

The World Database on Protected Areas (WDPA) is reviewed in Part I or Chapter 5 providing insight into protected area coverage and representation. The proportion of the world's surface under some form of protection is not insignificant: The database indicates that 14.6% (19.6 million km^2) of the world's terrestrial area, excluding the Antarctica, and 2.2% (6 million km^2) of the global ocean are managed. However, that protected area portfolio does not protect the full range of species and ecological diversity: Of the 823 terrestrial ecoregions and 232 marine ecoregions that have been defined (Olson et al., 2001; Spalding et al., 2007), only about one third of the terrestrial ecoregions meet the Convention on Biological Diversity's (CBD) 17% coverage target, and 13% of the marine ecoregions meet the 10% target. In 2010, 10% of the terrestrial ecoregions had less than 1% of their area protected, and 59% of the 232 marine ecoregions had less than 1% protected (Bertzky et al., 2012). The current protected area system does not meet the targets for 2020, especially in terms of representation, even taking into account that our scientific understanding of the patterns of biodiversity is rudimentary.

Part IV explores how the current network of protected areas might best be expanded to ensure that the portfolio is more ecologically representative and covers areas of particular importance to biodiversity. Chapter 12 introduces the concept of Important Bird Areas (IBAs) and Alliance for Zero Extinction (AZEs) sites. IBAs are places of international significance for the conservation of birds, and AZEs represent locations at which species extinctions are imminent unless the areas are appropriately protected. IBAs and AZEs have been located worldwide, and 10,993 IBAs and 588 AZEs have been identified. In 2010, only 28% of IBAs were completely contained within protected areas, while 49% were completely outside. Only 22% of AZEs were completely contained within protected areas, and 51% were found outside of these areas. In the case of both IBAs and AZEs, the calculated risk of extinction of species was higher outside of protected areas. If all unprotected or partially protected IBAs and AZEs were added to the current protected area network, this would necessitate an increase in the global network of 4.6 million km^2 and an increase of 17.5% in the protection of terrestrial habitat.

Chapter 4 examines what it would take to expand the protected area network to protect threatened mammal, bird and amphibian species. How much additional land would need to be protected to ensure the persistence of all threatened mammals, birds and amphibians? The current protected area network only met the persistence targets for 12.7% of the threatened species, but the persistence targets for all threatened mammals, birds and amphibians could be met if the global protected area system was strategically expanded to cover an additional 18.4% of the earth's surface.

From the analyses presented in these chapters, it is clear that even if Aichi Target 11 is met and 17% of the terrestrial environment and 10% of the oceans are protected, it would be insufficient to protect the world's threatened species. It is important to recognise the sampling bias in these analyses, as they are based on a subset of the vertebrate data, representing less than 2.5% of the world's described species.

Part III highlights a number of the drivers that will likely play an increasingly important role in determining future patterns of biodiversity loss. These drivers include land-use change, climate change and the exploitation of high value species. Chapter 9 focuses on examining changing land uses around national parks in the United States and the implications for biodiversity within the parks. Lands surrounding 48 parks strongly connected to the ecological functioning of each were delineated and defined as protected area-centred ecosystems (PACEs). Land-use change and climate change are then summarised over the recent past and projected into the future. A take-home lesson was the great variability across the 48 PACEs in their vulnerability to land-use change and climate change. This approach has great utility in identifying landscapes that will be particularly vulnerable to future change and could greatly assist with adaptive management strategies.

Chapter 11 examines the importance of protected areas in conserving commercially high value (CHV) species found in Asia. CHV species analysed include:

- Tigers (*Panthera tigris*)
- Asian elephants (*Elephas maximus*)
- Asian rhinoceros species
  - Greater one-horned rhinoceros (*Rhinoceros unicornis*)
  - Javan rhinoceros (*Rhinoceros sondaicus*)
  - Sumatran rhinoceros (*Dicerorhinus sumatrensis*)

- Asian wild cattle species
  - Banteng (*Bos javanicus*)
  - Gaur (*Bos gaurus*)

In all of these cases, as populations have been exploited and species ranges have been restricted, protected areas provide the last refuges for these species. With the high value of these species, especially in illegal markets, the challenge is securing adequate funding to protect them. For example, effective tiger conservation is estimated to cost roughly US$930 per km^2 per year (Walston et al., 2010), but available funding is in the range of US$500 per km^2 per year. Nevertheless, criteria can be identified that would allow effective management of CHV species.

Part II highlights one of the greatest challenges to the conservation movement – how to effectively monitor species, ecosystems and the drivers of biodiversity loss. Without this information, it is difficult to empirically demonstrate which interventions are working or which conservation sites are more successful than others at safeguarding biodiversity. New insights, methods and approaches are beginning to transform the field of conservation. Chapter 6 examines available vertebrate population time series data for protected areas; trends are assessed in African protected areas, and future scenarios are modelled to explore the relative importance of management and protected area expansion. Data for these analyses were based on some 4337 populations of 1543 species in 977 protected areas, an extensive collection but one that represents less than 1% of the protected areas listed in the WDPA. Regions with the greatest data gaps include South America and Southeast Asia. The case study from Africa demonstrates that in the sampled protected areas, the number of large mammal species may have decreased by half since 1970. The analysis identifies major differences in regional trends. What are the factors that account for these trends? One lesson is that the area under protection is frequently less important than the effectiveness of management. Expending efforts and budgets on management might frequently be more effective than seeking to expand the protected area portfolio. Business as usual is not a good long-term strategy. This chapter demonstrates that though data are still limited, especially from some of the most biodiverse parts of the planet, aggregating population time series data shows great promise for providing conservationists with robust trend data to help with management from the local to global scale.

Chapter 13 reviews approaches to monitoring species in protected areas. Camera trapping is identified as one of the most scientifically robust and cost-effective approaches for monitoring diurnal, nocturnal and crepuscular medium- to large-sized vertebrates. This approach is rapidly expanding and helping to provide insight into abundance, distribution and species richness in some of the most data-poor parts of the world. This data can be aggregated to provide trends in abundance from the site level to global scale and will greatly improve the current species time series datasets.

Finally, Chapter 14, introduces the latest science on monitoring biodiversity using satellite technology to guide protected area design and management. The multiple uses of satellite technology to help manage protected areas are explored: monitoring populations of specific species (e.g. penguins to invasive species), recording land-use changes, relating climatic anomalies and measuring ecosystem services such as productivity. The application of satellites to monitor marine protected areas is highlighted: Satellites can provide information on surface temperature, salinity, sea surface height, wind speed and direction. In addition, they can detect threats to marine protected areas such as oil spills or illegal fishing vessels. While much of this technology is still in its infancy, satellites will provide major breakthroughs in monitoring and will revolutionise protected area monitoring as high-resolution data becomes more widely available, especially to developing countries.

Part II examines how the effectiveness of management in protected areas can be improved. Chapter 8 explores the drivers of species population trends in protected areas. This requires that in addition to monitoring biological outcomes (population trends), inputs, actions and context are also measured. They found that protected areas with information on both species population trends and data such as expenditure, capacity, threat intensity, specific interventions, ecological context, sociopolitical context or total area were extremely rare, making it difficult to empirically demonstrate conditions or actions that lead to successful conservation outcomes. Case studies and expert opinion surveys are also reviewed, demonstrating that factors driving species population trends in protected areas are complex but that management quality and resources are unsurprisingly extremely important factors.

Different protected area types differ in their effectiveness at protecting biodiversity. Chapter 2 reviews the categories of protected areas based on management regime (from strict reserves to management resource protected areas) and based on governance (from those run by government to those managed privately or by indigenous peoples/local communities). Each category offers options that may help improve upon existing approaches. One conclusion is that the traditional focus of studies on conservation effectiveness has been on state-run protected areas. This has resulted in a limited understanding of the effectiveness of other forms of protected area governance. This is addressed to some extent in Chapter 10 where the conservation effectiveness, as measured by biological indicators, of protected areas run by indigenous peoples or local communities is compared to that of strictly protected, government-run areas. Protected areas run by indigenous peoples or local communities represent 9.3% of all protected areas with a known governance type (Bertzky et al., 2012), so assessing their effectiveness in conserving biodiversity is critical. While no general patterns in terms of forest cover, deforestation rates and species diversity/richness were discernible, there were clear differences in the composition of vertebrate species under the two regimes. Not surprisingly, where certain species were used by local people, this influenced their viability. Biological diversity in community-managed forests also tended to be degraded over time. Studies also found that negative biological outcomes were associated with rapid economic development, population pressure, market incentives for conservation, conflicts and accessibility. There is clearly great variation among community-managed protected areas in their effectiveness at conserving biological diversity, but these areas are a critical part of the global protected area network and deserve greater research and attention.

Chapter 7 examines the effectiveness of the European national park network and Natura 2000 sites in conserving the three largest carnivores in Europe: the lynx (*Lynx lynx*), wolf (*Canis lupus*) and brown bear (*Ursus arctos*). The long-term viability for each of these species was assessed under different protected area network scenarios in countries across Europe. One conclusion is that in isolation, few countries in Europe can protect one or more viable population of large carnivores by themselves. Conservation planning for these species can only be effective if planning were done at the continental scale. While the Natura 2000 network was not designed to operate at this scale and led to many uncoordinated national networks, large carnivores in Europe were resilient to this design flaw, largely because of their ability to adapt to the matrix landscape outside protected areas. Suitable habitat outside of protected areas has been critical to their long-term survival, and populations of large carnivores in Europe are increasing.

Part I extends the argument for protected areas beyond biodiversity conservation. Protected areas are an important investment for society, and there are a number of ancillary arguments as to why they should be financed. Chapter 3 outlines the biodiversity and humanitarian implications of climate change and highlights the fundamental importance of protected areas in storing carbon. A conservative estimate is that protected areas globally store 312 billion tons of carbon or 15% of the terrestrial carbon stock. Ensuring that these carbon stocks are secured is in the interest of global health. Protected areas also play a critical role in climate change adaptation, helping to reduce the vulnerability of local communities by protecting watersheds and soil, maintaining features such as mangroves that shield communities from major tidal surges or helping to maintain food sources such as fisheries or other wild crop relatives. Protected areas are also responsible for providing a significant amount of drinking water to major cities such as New York, Tokyo, Sydney and Mumbai (Dudley & Stolton, 2003) and providing major sources of water for crop irrigation. In addition, they play a more traditional role of providing jobs through ecotourism.

Recent estimates of the costs of supporting an effectively managed and representative global protected area network range from US$34 to US$79 billion per year (Butchart et al., 2012; McCarthy et al., 2012). Current expenditures are likely closer to US$6 billion (Balmford & Whitten, 2003). However, expenditures of even US$79 billion are comparably inexpensive when compared to the estimated annual value for goods and services provided by the global protected area system, which are closer to US$4400–5200 billion (Balmford et al., 2002). The societal justification for establishing and managing many protected areas has frequently been their value in conserving biodiversity, but clearly the value of this land use needs to be quantified in terms of economic benefits and development agendas. We need to better understand and communicate the economic and social value of each protected area and the larger national and global networks.

The future of protected areas

So what does the future look like for the world's protected areas? First, it is clear that even if we meet Target 11 by 2020 and 17% of the land and inland waters and 10% of the ocean are protected, this will be insufficient to ensure viable populations of species that are currently known to be threatened. We would therefore not meet Target 12 of the CBD 2020 targets, which calls for the '…extinction of known threatened species to be prevented and their conservation status, particularly of those most in decline, to be improved and sustained by 2020' (CBD, 2010). The global community recognises that the numbers specified in Target 11 have no scientific or ecological basis (see Chapter 1), and ultimately, there is no scientific answer to how much space we should leave for other forms of life. This is a moral and ethical discussion. Our decision on how much space to set aside for other forms of life, and where to do so, will have major implications for generations to come.

While the broad question of how much area should be set aside as protected areas is not scientifically tractable, the more specific question of how much should be set aside to prevent the extinction of individual species (Target 12) can be determined empirically. Chapter 6 begins to address this important question. At the very least, we should follow this approach to assess the area needs of threatened species and do this beyond mammals, birds and amphibians. Ecological representation can also be empirically measured (preliminary findings are presented in Chapter 4), but this too needs to be explored in much greater detail. The conservation community has a responsibility to identify the area needs for species. If this is not done, and a clear vision articulated and advocated, the future of many species, and perhaps our own, will be defined by what a few policymakers feel is politically appropriate or feasible.

An increase in protected areas is obviously positive for biodiversity, and countries should be held accountable to the commitments that have already been made under Target 12. However, countries are under no formal obligation to define a strategy of how the target will be met and the timeline for implementation. Civil society has a role in helping governments develop these strategies. National conservation non-governmental organisations (NGOs) should be demanding these implementation strategies and encouraging countries to report progress against the intended timeline.

In Part IV, it is evident there is a biodiversity monitoring revolution in play. The increased capacity to monitor biodiversity will completely transform the field of conservation with a particular impact on protected area management. With new technology such as remote monitoring units and satellites, protected area managers will soon be able to cost-effectively measure trends in species and ecosystems as well as drivers of biodiversity loss. It will be possible to disaggregate this information to explore trends by protected area, major ecosystem, region and county or even at the global scale. Conservation will finally join other disciplines in having real-time biodiversity indicators. We will then be able to clearly state which protected areas are truly safeguarding biodiversity and which interventions are having the intended impact, allowing us to rapidly scale up activities that truly work. This technology will revolutionise surveillance in protected areas and public engagement.

We currently have protected areas globally that contain extremely economically valuable species, yet as described in Chapter 11, many lack the resources to effectively protect them. It is much like having an art gallery that lacks an alarm system or sufficient security staff; thieves can simply walk off with the most expensive pieces. New technology will soon help effectively provide an alarm system for the world's protected areas, but much more than an early warning system is needed to secure the future of CHV species. The conservation community needs to develop common standards to effectively protect species. CHV species such as rhino, elephant and tigers are but the most visible manifestation of the needs to protect valuable biodiversity. For instance, a new standard should include the recommendations from Chapter 11 as well as a common platform for patrol-based monitoring such as the Self-Monitoring, Analysis and Reporting Technology (SMART) system and a clear community engagement strategy. Sufficient funding then needs to be raised to rapidly roll out these new standards across the world's protected areas.

Chapter 2 defines many different management categories that have been adopted by the International Union for Conservation of Nature. One common goal of protected areas is the protection of biodiversity. The goal of protection however should not be restricted to protected areas, and the conservation of biodiversity must also be a goal in the broader landscape and seascape matrix. It may be more appropriate to think in terms of zoning, with some land and sea use zones being more wildlife friendly than others. Ensuring that this broader landscape matrix is maintained through effective zoning is essential for the long-term persistence of species such as the large European carnivores reviewed in Chapter 7.

Engaging the public and making the social and economic argument for protected areas should be a continued effort of the NGO and scientific community. Technology is transforming the way people engage with protected areas. A mobile application, Instant Wild, now exists sending hundreds of thousands of people images of amazing species from protected areas all over the world, asking them to get involved in the identification process (ZSL, 2014). Soon, many of the world's protected areas will be brought to life through digital media reaching millions. People will be able to track individual species or watch predator–prey interactions from their desktop or smartphone. They may even have the option to adopt a particular square of a protected area that they can also help to monitor and protect. While digital engagement in a world that's population is predominantly urban is extremely important, nothing will replace, encouraging the next generation to experience nature first hand. Initiatives bringing local children to protected areas should be part of national curriculums and a fundamental activity of local NGOs. If we fail to communicate the importance of protected areas to the next generation, they will be quickly lost as land-use trade-offs intensify.

The conservation of biodiversity in protected areas is rooted in an ethical argument. It is our responsibility to protect the roughly 8.7 million other forms of life because they have a right to exist and future generations have a right to live in a world that contains the great diversity of life. This is fundamentally the reason that most people want to protect species, because they think it is the right thing to do. However, we also need to support this with strong economic and human development arguments that are well articulated in Chapter 1. The conservation community needs a much stronger research agenda that focuses on how to connect the next generation with nature and how to best measure and articulate the economic and social values of protected areas.

With increasing population growth and associated pressures for agricultural expansion and resource extraction, there is going to be more and more pressure to both develop and extract resources from the world's protected areas. For example, there are oil and gas concessions that are inside 27% of the natural World Heritage Sites, and although none of these currently have active oil wells (Turner, 2012), companies that hold these concessions are expecting to drill. What is permissible within protected areas? What activities will not be tolerated in protected areas under specific management regimes? When most of the world is effectively zoned for agriculture or resource extraction, do we really need to exploit the 14.6% of land and 2.2% of the oceans that have been set aside for nature? The conservation community needs to develop a clear position and work with government and industry to ensure they make lasting commitments to this position.

Finally, much better reporting on the effectiveness of the world's protected areas is needed, clearly defining progress towards agreed targets as well as state, pressure and response monitoring of protected areas from the site to national to global scale. This document would be fundamental in the continual evaluation and planning process of the global protected area network. It is hoped that this book will lay the foundation for such a report and the effective planning of the global protected area network as we start to look well beyond 2020.

References

Balmford, A. et al., 2002. Economic reasons for conserving wild nature. *Science*, 297(5583), p. 950–953.

Balmford, A. & Whitten, T., 2003. Who should pay for tropical conservation and how could the costs be met. *Oryx*, 37, p. 238–250.

Bertzky, B. et al., 2012. *Protected planet report 2012: Tracking progress towards global targets for protected areas*, Gland & Cambridge: IUCN & UNEP-WCMC.

Bruner, A. G., Gullison, R. E., Rice, R. E. & da Fonseca, G. A. B., 2001. Effectiveness of parks in protecting tropical biodiversity. *Science*, 291(125), p. 125–128.

Butchart, S. H. M. et al., 2010. Global biodiversity: Indicators of recent declines. *Science*, 328(5982), p. 1164–1168.

Butchart, S. H. M. et al., 2012. Protecting important sites for biodiversity contributes to meeting global conservation targets. *PLoS ONE*, 7(3), p. e32529.

CBD, 2010. *Decisions adopted by the CoP to the CBD at its 10th meeting (UNEP/CBD/COP/DEC/X/2)*, Montreal: Secretariat of the CBD.

Dudley, N. & Stolton, S., 2003. *Running pure: The importance of forest protected areas to drinking water*. Gland: WWF.

Emmott, S., 2013. *Ten billion*. New York: Vintage.

McCarthy, D. P. et al., 2012. Financial costs of meeting global biodiversity conservation targets: Current spending and unmet needs. *Science*, 338(6109), p. 946–949.

Olson, D. M. et al., 2001. Terrestrial ecoregions of the world: A new map of life on earth. *BioScience*, 51(11), p. 933–938.

Selig, E. R. & Bruno, J. F., 2010. A global analysis of the effectiveness of marine protected areas in preventing coral loss. *PLoS ONE*, 5(2), e9278.

Spalding, M. D. et al., 2007. Marine ecoregions of the world: A bioregionalization of coastal and shelf areas. *BioScience*, 57(7), p. 573–583.

Turner, S. D., 2012. World heritage sites and the extractive industries: An independent study commissioned by IUCN in conjunction with the UNESCO World Heritage Centre, ICCM & Shell. [Online] Available at: http://cmsdata.iucn.org/downloads/whs_and_extractive_industries_20_jun_12.pdf [Accessed 22 July 2015].

United Nations, 2013. *World population prospects: The 2012 revision, key findings and advance tables*, New York: Department of Economic and Social Affairs, Population Division.

Walston, J. et al., 2010. Bringing the tiger back from the brink: The six percent solution. *PLoS Biology*, 8(9), e1000485.

WDPA, 2014. *The world database on protected areas (WDPA)*, Gland & Cambridge: IUCN & UNEP-WCMC.

WWF, 2012. *Living planet report 2012*, Gland: WWF.

ZSL, 2014. *InstantWild*, London: Zoological Society of London.

Part I
The Global Protected Area Portfolio

1

Government Commitments for Protected Areas: Status of Implementation and Sources of Leverage to Enhance Ambition

L. Krueger

The Nature Conservancy, Arlington, VA, USA

Introduction

The past 20 years has seen a dramatic increase in the number and extent of protected areas worldwide. Although the legal creation of a park does not guarantee the conservation of the biodiversity within it, it is usually an essential first step for securing natural values in a place, particularly as global population and development pressures leave few landscapes and seascapes untouched by humanity. As such, the ongoing creation of protected areas and expansion of the protected area portfolio are often cited as some of the greatest successes of conservation. While the establishment of protected areas clearly recognizes the value of protection as a way to mitigate human impacts on biodiversity, it is appropriate to review the political mechanisms that have driven these achievements: What is the role of formal political commitments, and what are the potentials and limitations as we seek more comprehensive and effective conservation outcomes moving forward?

Governments at many levels have long recognized the value of protected areas, and indeed, the very concept of conservation or reserve areas has an ancient provenance. But since the 1992 Rio Summit, the global protected area estate has dramatically expanded largely in response to explicit commitments made by governments in the international fora. International treaties have contributed to a process of changing global norms and have encouraged governments to make deeper commitments to protected areas; among them are the World Heritage Convention, Convention on Biological Diversity (CBD), and the Ramsar Convention on Wetlands of International Importance. In addition, the

Protected Areas: Are They Safeguarding Biodiversity?, First Edition.
Edited by Lucas N. Joppa, Jonathan E. M. Baillie and John G. Robinson.

International Union for the Conservation of Nature (IUCN) World Commission on Protected Areas (WCPA), while not a formal intergovernmental agreement, has done much to promote a global community of practice in support of protected areas. Most recently, in 2010, 193 nations in the world committed to the CBD's Aichi Biodiversity Target 11 to increase "effectively and equitably managed, ecologically representative and well connected systems of protected areas" to at least 17% of the terrestrial and inland water and 10% of the coastal and marine areas by 2020 (CBD, 2010a).

This chapter reviews the role and status of legal frameworks and other commitments for protected areas, and it explores the relationship between scientific evidence and political practicality in implementing current targets and achieving the more ambitious ones. The rationale for these targets is contested. On the one hand, they are seen as underambitious, as biodiversity research has demonstrated that even successful implementation of current targets is unlikely to prevent unprecedented levels of biodiversity loss. On the other hand, the very concept of protected areas is challenged in some quarters as outmoded, and their expansion is seen as a hindrance to more economically profitable land uses. Under these circumstances, the international policy debates around protected areas become crucial arenas for reconciling multiple societal goals and can help the world achieve a rational and effective level of protection given our best understanding of the science and the costs and benefits of alternatives. Although the link between international political commitment and action is not always direct, it can establish channels to promote deeper and more sustainable public support for conservation.

Emergence and evolution of government commitments on protected areas

Protected areas are not a new concept. Rulers have made formal or informal declarations to protect areas for centuries, usually to protect royal hunting grounds or sacred sites. Probably the oldest continuously protected area still managed as such today is the Bogd Khan Uul park in Mongolia, established by Buddhist and Manchu authorities in 1778 and currently managed as a strictly protected area and a UNESCO Biosphere Reserve (UNESCO, 2013). The creation of Yellowstone by the United States Congress in 1872 is often recognized as the inception of the modern era of formally designated protected areas, but it hardly represented a watershed event: National park creation continued to be sporadic and modest for most of the next hundred years after Yellowstone.

One of the earliest efforts to build an international consensus around protected areas was the International Conference for the Protection of Fauna and Flora, held in London in 1933. The conference led to the first attempts to categorize types of protected areas (Phillips, 2004). It was not until 1962, however, that the international community assembled its first UN list of parks for the First World Conference on

National Parks, held in Seattle. The Seattle conference was bedeviled by debates over terminology and nomenclature in deciding what could and could not be included on the United Nations list. These debates stimulated further attempts to define management categories through the International Commission on National Parks (later the WCPA, a voluntary commission of the IUCN).

Though IUCN is not a legally binding treaty mechanism, its voluntary approaches have been essential in building a community of practice around protected areas and spurring their use as a tool by governments. WCPA's continuing efforts to update and refine the protected area management categories (in 1978, 1994, and 2012) contributed to enhanced national efforts to build national systems by providing a template that allowed the categorization of national laws and helping standardize nomenclature (Dillon, 2004). The CBD further promoted this effort in 2004 by formally endorsing the IUCN categories and encouraging governments to assign categories consistent with those developed by IUCN (CBD, 2004).

These efforts by IUCN and national protected area managers established a consistent definition of protected areas, advanced the understanding of protected area characteristics, and addressed the need for consistent international standards. However, none of these works bound governments to specific commitments or obligations. A series of more formal international agreements starting in the 1970s helped disseminate emerging scientific consensus about environmental problems among policy makers and began to create a web of obligations on nations that have steadily grown in scope and specificity (Table 1.1).

The Man and Biosphere Conference held in Paris in 1968 was credited as being the first intergovernmental meeting to recognize the transboundary nature of many environmental problems. Organized by UNESCO in collaboration with Food and Agricultural Organization (FAO) of the United States and the World Health Organization (WHO), and in cooperation with the International Biological Programme (IBP) and IUCN, this event, attended by representatives from 60 countries, was the first worldwide meeting on global environmental issues. The Man and the Biosphere Programme (MAB) and the concept of biosphere reserves were created in part to encourage the creation of a global network of protected areas (Dyer & Holland, 1988).

A few years later, the UN Conference on the Human Environment (the Stockholm Conference) represented the first attempt to forge a global agenda to address global environmental problems. Though the documentary products of this conference largely contained broad policy goals rather than specific objectives, it had a profound impact in drawing political and public attention to the natural environment and led directly to subsequent agreements that consolidated the role of protected areas as a tool for conservation. Both the World Heritage Convention and the Ramsar Convention on Wetlands of International Importance, negotiated around the same time as Stockholm, required signatories to designate sites for protection and further catalyzed the normative development of protected areas as a focus of conservation efforts even beyond what was legally called for

Table 1.1 **Some international milestones in the evolution of global protected area commitments**

Year	Event	Outcome/target related to protected areas
1968	Biosphere Conference	Led to creation of Man and Biosphere Programme (in 1971) and UNESCO Biosphere Reserves (1974)
1972	World Heritage Convention	Conserve "precisely delineated areas which constitute the habitat of threatened species of animals and plants of outstanding universal value from the point of view of science or conservation..."
1972	UN Conference on the Human Environment	Stockholm declaration calls for safeguarding "representative samples of natural ecosystems..."
1975	Ramsar Convention enters into force	Parties must designate suitable wetlands for the List of Wetlands of International Importance ("Ramsar List") and ensure their effective management
1982	3rd World Parks Congress	Objective to protect 10% of the terrestrial ecosystems
1987	Brundtland Commission	Called on countries to "triple" extent of protected areas (to ~12%)
1992	4th World Parks Congress	>10% of each major biome (by 2000)
1992	CBD	Calls for establishment of national systems of protected areas
2002	CBD COP 6	10% of the world's ecological regions protected (Decision VI/9)
2002	World Summit on Sustainable Development	First Marine Protected Areas target: MPA networks established by 2012
2004	CBD COP 7	10% (+ effective, comprehensive, representative qualifiers) with time-bound milestones (Decision VII/28)
2010	CBD COP 10	Strategic Plan/Aichi Target 11: 17% terrestrial/10% marine ecosystems protected by 2020

in either convention. Indeed, protected areas have become so ingrained as a conservation tool that they have been frequently used as a proxy for biodiversity conservation effort overall (Chape et al., 2005).

This timeline of increasing political action and commitment parallels the dramatic surge in the number and extent of protected areas declared from the 1970s onward (Figure 1.1). While it is difficult to assign causality to the complex societal decisions that lead to protected area creation, the very fact that protected area extent can be easily measured and reported would make it more likely to be

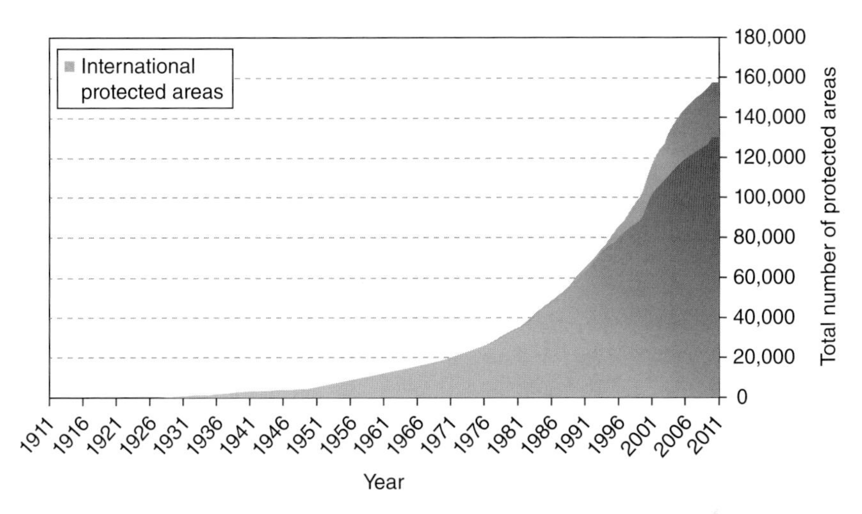

Figure 1.1 **Growth in number of nationally and internationally designated protected areas (1911–2011). Source: Data from WDPA (2010)**

implemented. As discussed below in relation to the CBD's increasing specificity of targets on protected areas, the successful implementation of any international commitment typically requires clear definition of requirements, targets, and milestones. Indeed, the past four decades of international discourse on protected areas has provided the tools that have led to the important achievements thus far.

The creation and ongoing expansion of protected areas in virtually all countries over the past 40 years have arguably been some of the greatest success stories in conservation. Overall, protected areas have become one of the most predominant land use categories on Earth (Chape et al., 2005). As of 2013, the World Database on Protected Areas (WDPA) listed over 205,000 designated protected areas, covering 12.7% of the global land area (outside Antarctica) and 7.2% of the coastal waters (0–12 nautical miles) (WDPA, 2010). Prompted by increasingly urgent scientific warnings on biodiversity loss and supported by an emerging international community of practice around protected areas, governments have been commendably responsive both through commitment and action in developing national protected area networks.

Role of the CBD and its Programme of Work on Protected Areas (2004)

The CBD, signed at the Rio Earth Summit in 1992, has gradually emerged as the most comprehensive legal framework for protected areas. With 193 national parties, the CBD has near-universal global membership.[1] Article 8 of the Convention recognizes protected areas as a key strategy for biodiversity conservation, but it originally provided little guidance or specificity for parties, beyond requiring each country to:

- Establish a national system of protected areas.
- Develop guidelines for their establishment and management.

This rather vague obligation, which in any case had been nominally met by most parties prior to their accession to the Convention, compelled little direct activity by countries in the initial years after the treaty's entry into force.

In fact, the CBD's weakness is that it was designed as a framework convention, much like the UN Framework Convention on Climate Change (which led to the Kyoto Protocol) or the Vienna Convention for the Protection of the Ozone Layer (which led to the Montreal Protocol). As such, it set out broad principles and did not impose true obligations on parties but rather was intended to be a springboard for subsequent protocols that would provide measurable targets and objectives obliging states to act. The CBD's work on protected areas has not yet resulted in a formal protocol, but an evolving series of negotiated decisions and reporting requirements related to protected areas increasingly bear many of the hallmarks of such a binding international agreement.

In 2004, at the 7th Conference of the Parties (COP 7) of the CBD, the first major decision on protected areas was adopted that substantially raised the bar on national commitments regarding protected areas. The decision, called the Programme of Work on Protected Areas (PoWPA), was an important attempt to create an implementation system by increasing the level of specificity of requirements, with implied obligations and responsibilities for parties to the Convention, including the adoption of time-bound interim targets for the achievement of a comprehensive and integrated system of protected areas. There are four elements of the PoWPA:

1. Establishing, strengthening, and managing protected area systems.
2. Strengthening governance, participation, equity, and benefit sharing.
3. Enabling activities/policy.
4. Fostering standards, assessment, and monitoring remain the building blocks of implementing national protected areas systems and reporting to the CBD on progress.

Notably, the PoWPA included a quantitative target that at least 10% of each country's terrestrial ecosystems be protected by 2010. The agreement to measurable time-bound goals was a significant achievement for a treaty that had few tools to encourage compliance at the time.

In truth, parties to the agreement have missed many of the PoWPA deadlines, but in the decade following COP 7, there has been a continuous effort to build on successes and strengthen protected area commitments through training, development assistance and cooperation, and exchange. PoWPA remains the framework for implementing protected area goals, although it has been supplemented by the Strategic Plan Targets, the Aichi Targets, adopted at the CBD's 10th Conference of the Parties (COP 10).

The CBD 2020 Strategic Plan and Aichi Target 11

The difficulties of measuring (much less achieving) progress toward the Convention's overall 2010 target[2] and the perceived success of the PoWPA model of cooperative progress toward specific goals led to a determination among many negotiators at the COP 10 (in 2010) to push for more measurable targets as part of the CBD's new strategy for the decade 2011–2020. The NGO community and many negotiators made frequent demands for targets that were *S*pecific, *M*easurable, *A*ttainable, *R*elevant, and *T*ime sensitive (SMART). The resulting targets, called the Aichi Targets after the prefecture in Japan in which they were negotiated, included, among other numerical goals, a sizable increase in the extent of terrestrial protected areas to be achieved by 2020. Target 11 calls for at least 17% of the terrestrial and 10% of the marine ecosystems to be protected. It also includes important modifiers for these numerical targets, such as the requirements that protected areas cover key biodiversity areas and that they be areas important for ecosystem services, representative of ecosystems, well connected and linked to other area-based conservation measures, and integrated into wider landscapes and seascapes.[3]

How did negotiators arrive at these numerical targets and what was the comparative influence of science versus politics in motivating them? As can be seen from some of the other chapters in this volume, the science is not clear on how much protection is necessary, and the answer varies considerably by site. The scientifically derived, qualitative goals of Target 11 (related to representativeness, ecosystem services, and connectivity, among others) are overlapping and provide wide latitude for conservation planning that either maximizes or minimizes the land needed for protection. Unsurprisingly, the target percentages proposed in the negotiations varied widely. Prior to the COP, the draft text prepared by the CBD Secretariat included both 15 and 20% terrestrial targets in brackets (CBD, 2010b). Some countries suggested over 30%, and others held fast to the 10% target from the 2004 PoWPA decision or would have preferred no target at all. Some felt the number was not important, except for the fact that there needed to be a number against which to measure progress, adhering to the maxim that specificity of goals improves implementation. The marine target of 10% was retained from the 2004 PoWPA agreement, after considerable debate at the COP 10 (Spalding et al., 2013). This reflected the considerable remaining distance between current marine protected area (MPA) coverage, 1.31% in 2010, when the Aichi Targets and 2004 targets were being negotiated.

With no clear scientific prescription and nations suggesting a wide range of targets and goals, it was largely up to negotiators to come up with numbers that were at least scientifically plausible, achievable in the eyes of most governments, and preferably more extensive than what had been proposed in the 2004 PoWPA decision. The 17% target was ultimately a split-the-difference compromise based on the figures that had predominated in the lead-up discussions to the COP. It was determined neither by science nor politics but with deference to both and with the very practical aim of spurring countries to greater action by setting an ambitious yet attainable target. Yet how meaningful was the adoption of this target?

The importance of national implementation in the context of the global target

Implementation of treaty obligations is important at multiple jurisdictions but no more so than at the national level. Only national governments can be accountable both to their international commitments and their own publics, be responsive to reporting requirements and international norms, and have the authority and legitimacy to make enforceable policy changes domestically.

Despite the specificity of many of its targets, the CBD Strategic Plan was designed to be a flexible framework that would allow nations considerable leeway in their own implementation actions. The global protected area target was relatively easy for countries to accept at the COP 10, because in doing so they were not actually taking a position on what they would do domestically. Each country is allowed to designate its own percentage of protection or need not establish a national numerical target at all (but all are urged to do so). While this legal loophole is important to countries to not jeopardize their formal compliance with the Convention, the target has nonetheless served as a benchmark against which both global and national performances are being evaluated.

For example, the 2012 Protected Planet Report provides a global analysis of current performance against the current global commitments (Bertzky et al., 2012). As the report notes, in terms of gross coverage, the world has already exceeded the previous target of 10% for terrestrial protected areas. The remaining gap to attain 17% of the terrestrial coverage requires an additional 5.5 million km^2. When compared with the approximately 17 million km^2 already under terrestrial protection, this goal seems well within reach. The 10% marine target is somewhat more challenging, with a need for an additional 9.7 million km^2, but is still numerically attainable (Spalding et al., 2013).

However, the situation looks more complex when one starts to incorporate the qualifiers in Aichi Target 11. Fewer than 50% of the world's 823 terrestrial ecoregions have close to adequate representation within the protected area network. In order to meet the target in terms of representativeness, an additional 10.8 million km^2 is required to fill the gap. Effective management also remains elusive. According to a recent study, less than 30% of the protected areas have a management plan, and the actual level of effective management is likely to be less than 25% of all protected areas (Leverington et al., 2010). Quantitative assessments of the other components of Target 11—equity, connectivity, and landscape integration—are scarce, but the Protected Planet Report points to signs of progress in these areas (Bertzky et al., 2012) (Figure 1.2).

This global and ecoregional assessment is certainly helpful for seeing the big picture, but any appraisal of whether CBD target setting is useful for spurring action must examine progress at the country level. In addition to its work on the Strategic Plan and Aichi Targets, COP 10 adopted a protected area decision that called parties to develop protected area action plans that would provide a road map for the completion of their national systems (COP 10, Decision 31). Overall, the proposed national implementation goals identified in these plans are impressively in line with the 17% terrestrial target, indicating that countries are using the global targets as a benchmark. It is important to note that in many cases, the action plans themselves do not carry legal force at the national level and often do not fully address the means to achieve the quality and effectiveness goals called for by both the PoWPA and Aichi Target 11. However, of the parties identifying quantifiable targets in their protected area action plans, 70 countries stated goals of 10–30%, with 46 parties setting their target at or above 17%. These new national targets will presumably address many of the shortcomings in ecological representativeness identified in earlier studies. National expansion plans and targets for protected areas are largely based on the scientific gap analyses that were conducted under the original PoWPA commitments (and supported with Global Environment Facility (GEF) funding) in 2005–2009 and are likely to adhere to that scientific guidance (Figure 1.3).

In the marine realm, the Aichi Target is unclear as to whether the target pertains to territorial waters (up to 12 nautical miles from the shore) or countries' exclusive economic zones (EEZs), the 200 nautical mile area within which governments have more limited jurisdiction but can control fishing and other economic activities that are the most immediate threat to marine biodiversity. Forty-four countries have specified protected area targets ranging from 3 to 15%. Seventy parties whose combined waters add up to nearly half of the world's territorial waters have proposed marine targets which would amount to over 11% of their combined territorial waters being protected. This would surpass the global marine and coastal target of 10% protection without taking ecological representativeness into account. Twenty-eight countries and territories have over 10% MPA coverage in place already (Spalding et al., 2013). Interestingly, several small island developing states have set ambitious MPA targets of 25% or more although collectively they currently only have 2.8% MPAs. These large gaps between proposed national targets and current MPA coverage may raise doubts about the feasibility of implementation and effective management for those areas. Particularly for protection in the EEZs, the challenges of governance and enforcement of restrictions

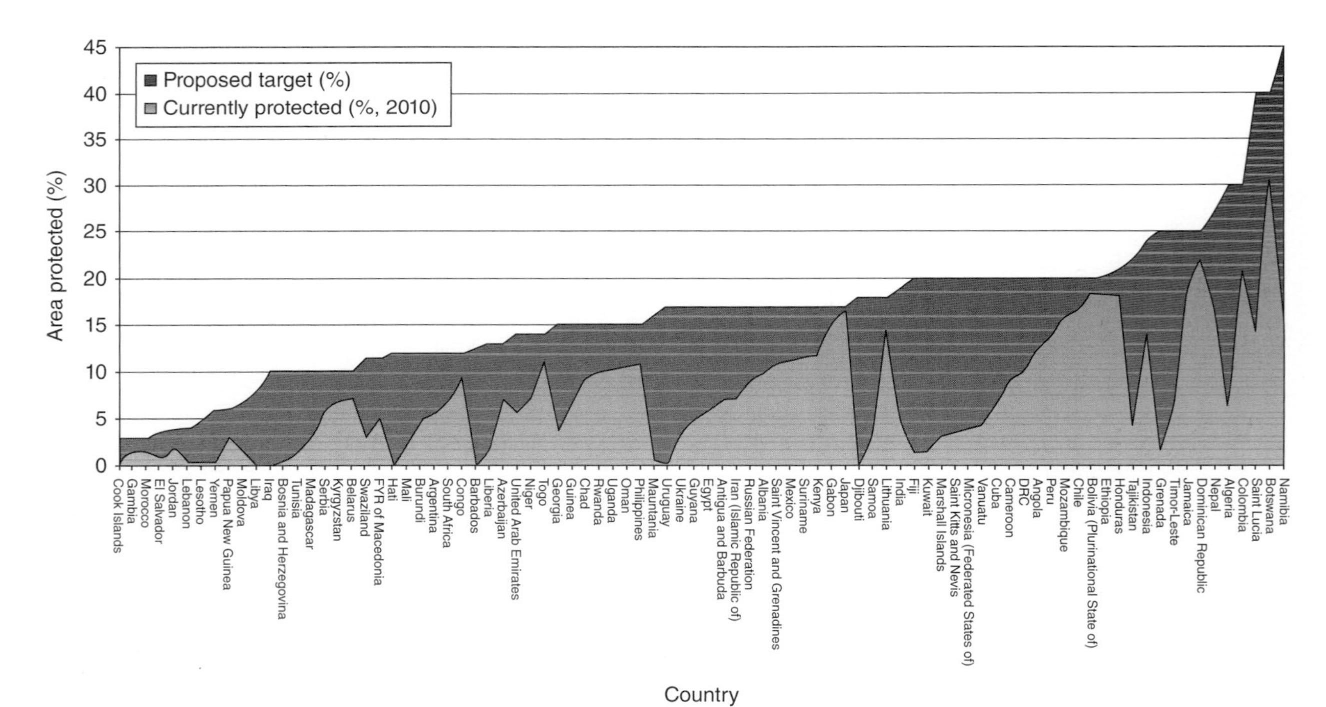

Figure 1.2 **Current and targeted percent of terrestrial area under protection for 86 countries. Source: Data from CBD (2012)**

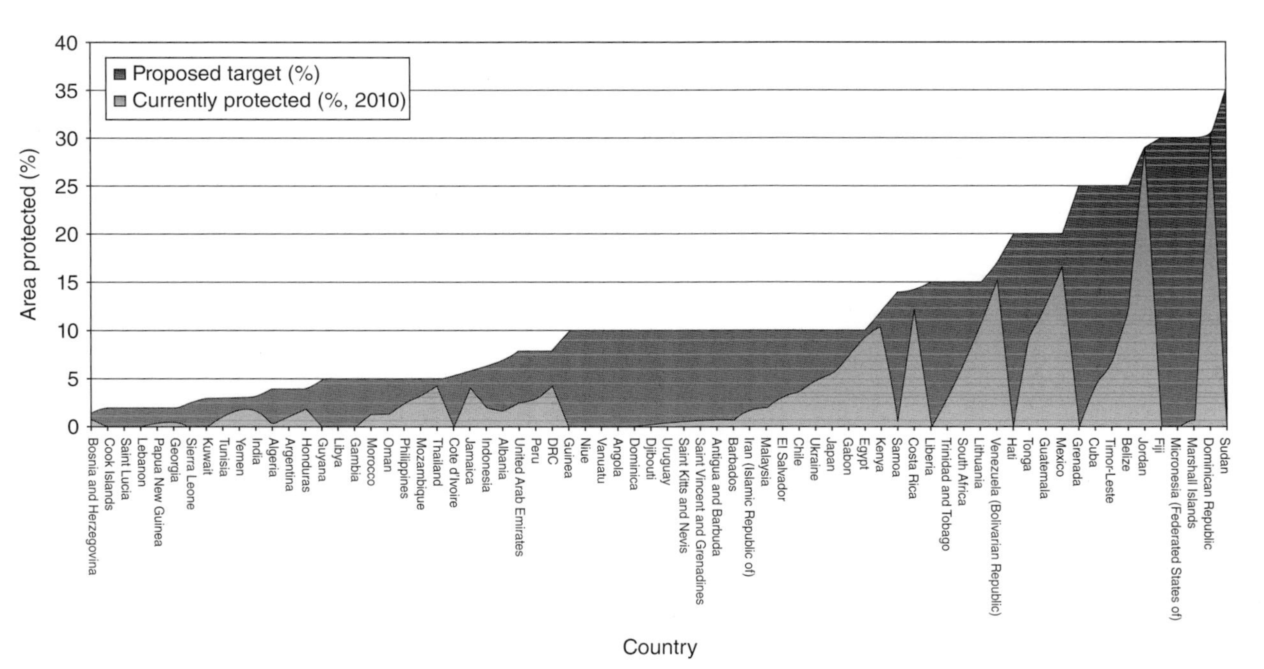

Figure 1.3 **Marine areas under protection and national MPA targets for 70 countries. Source: Data from CBD (2012)**

over such large areas are daunting for small countries with limited personnel and other resources.

Many of these national commitments are still unaccompanied by the necessary implementation, budgets, or legislation, and they are sometimes based on the expert views of a country's CBD focal points, reflecting that experts' best understanding of what is scientifically appropriate and politically attainable at the national level. Nonetheless, the progress and intentions outlined in countries' reports to the CBD demonstrate a strong interest to conform to the international standard established by Aichi Target 11 for both terrestrial and marine areas.

One gap that cannot be addressed by specific national commitments is the need for MPAs in areas beyond national jurisdiction (High Seas). The Aichi Marine Target is not explicitly limited to territorial waters, but as agreement on the High Seas involves all nations, it is generally recognized that any formal commitment to establish protected areas in these zones must be made by the UN Convention on the Law of the Sea (UNCLOS).[4] However, both the CBD, through its work on ecologically and biologically significant areas (EBSAs), and IUCN's Global Ocean Biodiversity Initiative (GOBI) have contributed a large body of scientific and technical input to identify areas that might be designated and, in doing so, have helped raise the political salience of the issue and increased the probability of further action. At the UN Conference on Sustainable Development or Rio +20 in 2012, global leaders committed to "address, on an urgent basis, the issue of the conservation and sustainable use of marine biological diversity of areas beyond national jurisdiction, including by taking a decision on the development of an international instrument under the United Nations Convention on the Law of the Sea" (United Nations, 2012). Although the task of achieving this vague global commitment will be exceptionally complex and time consuming for negotiators, it is being driven by a growing scientific consensus and technical support on the design of a High Seas protected area network.

Sources of progress toward Target 11

Progress toward developing a truly global network of protected areas has been aided by a number of institutional factors in the past decade:

- **Increasing specificity of requirements:** Part of what increases the effectiveness of political commitments is perceived precision of the obligations and the focus on monitoring and reporting of those obligations (Mitchell, 2011). The protected area action plans called for at the COP 10 (in decision X/31), combined with the more ambitious protected area targets in the new Strategic Plan, have given countries clear guidance on expectations. Within 15 months after the adoption of the action plan reporting requirement, over 105 governments had submitted plans, a remarkable level of compliance.

- **Increased alignment of donor funding:** Clearer goals have also provided donors with increased confidence that their support for protected areas meets well-defined priorities and contributes to global environmental goals. International support for implementation has been critical to raise the level of ambition and change norms defining what is desirable or achievable in terms of protection. Most significantly, Germany's initiation and funding of the LifeWeb initiative in support of expansion of the global protected area network has galvanized action in dozens of countries.
- **Capacity building:** The CBD Secretariat organized eight regional workshops between May 2011 and June 2012, designed to help countries share experiences and, in many cases, help draft the PoWPA action plans over the course of the workshops. A GEF umbrella grant directly supported national workshops to kick-start development and implementation of the PoWPA in over 40 countries.
- **Sustained focus:** The 2004 PoWPA still forms operative basis for protected area implementation globally, even though the targets from 2004 have been superseded by the targets under the new Strategic Plan adopted in 2010. The PoWPA continues to address all the conditional elements of Target 11 and more. Its specifications have stood the test of time, even as some of the deadlines have been missed, and it continues to provide guidance for the development of National Biodiversity Strategies and Action Plans (NBSAPs) and other implementation arrangements.

As the CBD Secretariat stated, "Simply put, focused action (emanating from goals and decisions) linked with available funding and structure capacity development leads to implementation on the ground" (CBD, 2012).

Strengthening the link between political commitment and action

International agreements, particularly environmental agreements, are difficult to enforce. They succeed largely by setting norms, creating a global community of practice, and requiring reporting and review systems that increase international scrutiny and hopefully encourage compliance. Nations are often viewed as complying with international environmental agreements out of a combination of self-interest (i.e., they implement what they would have implemented anyway in the absence of an agreement) and concern about potential reproach from the international community for lack of compliance (Chayes & Chayes, 1993; Downs et al., 1996). Assessing the value of international protected area instruments for driving change on the ground is particularly difficult in view of the fact that have almost universal adherence, which excludes the possibility of comparing the behaviors of signatories versus nonsignatories. In the near future, we may be able to use the action plans to evaluate the success of countries in achieving the targets they have set for themselves and determine whether target setting has an impact on outcomes, as other studies have suggested (Baettig et al., 2008). At present, these national-level commitments are too recent to be reflected in outcomes discussed in the fifth national reports to the CBD (due March 31, 2014); it will take several more years before their full merit can be judged.

What can be said now is that the activity promoted by PoWPA has achieved many less quantifiable objectives that are clearly important to building long-term political support for protected areas. Though initially just a negotiated CBD decision, the PoWPA evolved into a multistakeholder partnership, which helped strengthen the community of practice around protected areas in many countries, encouraged South–South cooperation, and spurred donor interest that elevated the cause of protected areas within government bureaucracies. Transnational activities organized under the PoWPA umbrella kept the focus on implementation through a bottom-up approach of building capacity and knowledge among those directly responsible for implementation at the national level. These activities created a vital bridge between the high-level commitments negotiated in an international forum (i.e., the CBD COP) and the real work that was taking place on the ground. Strengthening domestic constituencies around protected areas further helps governments clarify their interests and thus participate with greater confidence in subsequent cooperative efforts and discussions. These all provide important foundations for improving implementation of commitments made thus far.

Still, significant opportunities to reinforce progress and incentivize further action exist. As a first step, the protected area targets and subtargets must be well incorporated and justified as part of the new cross-sectoral NBSAPs that are required by the new CBD strategy. The CBD Strategic Plan decision requires all countries to revise their NBSAPs to improve alignment with the Aichi Targets and develop national targets (including protected area targets) that constitute their contribution to the CBD's strategy, revising their NBSAPs accordingly. The subsequent protected area decision from the COP 11 (2012) in Hyderabad also called on parties to integrate national action plans for various work programs (such as the action plans developed under PoWPA) into updated NBSAPs; to undertake major efforts, with appropriate support to achieve all elements of Aichi Target 11 to improve specifically MPAs in areas within their jurisdiction; and to attain those goals of the PoWPA that are lagging behind (UNEP/CBD/COP/11/L.9).

Integration of the protected area action plans into NBSAPs is not a straightforward matter of inserting protected areas into an NBSAP framework, as the NBSAPs will have to respond to an interlocking set of Aichi Targets. Protected areas may be a strategy to deal with several of the targets, but evaluating the relative value of protection as opposed to other sectoral mainstreaming or policy approaches may involve a complex cost–benefit calculation, not to mention a heavy dose of politics.

While the previous set of NBSAPs drafted in 2002–2003 rarely were applied as effective policy setting documents, the new NBSAPs are intended to drive what needs to be done on a number of fronts and serve as the primary framework of action for implementation and the basis for accessing national budgets, as well as bilateral, multilateral, and other financial supports from donors. Continued capacity building and transnational engagement will also be vital: The NBSAP Forum launched in 2013 is designed to bring together a large community of stakeholders, donors, subject matter experts, and practitioners to support ongoing NBSAP development and assist the GEF-financed NBSAP revisions being managed in approximately 120 countries by the United Nations Environment Programme (UNEP) and the United Nations Development Programme (UNDP). This will be way of integrating protected areas into the conservation blueprint and should also form the basis for future funding.

Overcoming funding and policy barriers to implementation of protected area commitments

Successful implementation of protected area commitments is not just a matter of political will: It also requires that real resource needs and policy gaps be addressed. The Aichi Targets and even the protected area targets specified in national plans were devised without a clear plan for paying for them. The numbers are indeed daunting: A recent analysis of the total cost of achieving Aichi Target 11 was estimated at US$270 billion, or US$33.75 billion per year between 2013 and 2020 (Ervin & Gidda, 2012). While the CBD is now working on its strategy for resource mobilization (itself the subject of a separate Aichi Target), there is widespread acknowledgment that few additional resources for protected areas are likely to come from direct budget allocations or traditional development assistance. This has spurred the search for more innovative finance mechanisms, such as payments for ecosystem services, compensation schemes that require investment in protection to offset biodiversity losses elsewhere, and addressing perverse incentives that harm protected areas and increase the cost of their establishment and maintenance. Targeted research is needed to measure the impacts of perverse subsidies on protected areas to shift the balance in government spending priorities that favor subsidies over meeting the comparatively modest funding requirements for effective conservation.

Addressing the finance gap is not just a budgeting problem. Scientific efforts must acknowledge and help resolve resource needs through better information about the cumulative impacts of human activities on biodiversity across the landscape. Information about the costs and benefits of various policy alternatives are often the most egregious of information gaps and the most significant barriers to conservation action. Assessing these trade-offs will require large landscape scale planning that integrates ecological values with competing production values that may exist in the landscape. While the decisions about land use will ultimately be political, more science is needed to characterize and quantify the costs and benefits.

Specific policy advances would also help countries gain support for protected areas and help them achieve targets in a more flexible and cost-effective manner. As described by Rao et al. (in this volume, Chapter 10), incorporating the recognition of other area-based conservation measures (such as indigenous and community conserved areas (ICCAs) and private protected areas (PPAs)) into the protected system could be a relatively quick route for many countries to achieving the numerical goals of the target. Indigenous groups have been protected sacred areas and species for centuries and continue to do so even without formal protected area status. ICCAs are recognized in the PoWPA as providing an important part of achieving protected area goals. Estimates range widely about the extent of areas under community management and the extent to which these areas contribute to biodiversity, but the 2011 WDPA includes over 1.1 million km^2 of ICCAs at 700 sites, which represents only a small fraction of the more than 300 million km^2 estimated to be owned and administered by communities globally (WDPA, 2010). As was discussed by governments at the COP 11 in Hyderabad, India, in 2012, if the effectiveness of ICCAs can be better assessed and attributed, their recognition and inclusion in the protected area estate could help double the extent of the area effectively conserved.

Similarly, a huge share of biodiversity exists on private lands, such as those managed by individual landowners, nonprofits, or corporations. What are the opportunities to increase coverage, connectivity, and overall effectiveness of protected areas through PPAs? Few countries have laws on the books that allow for PPAs, and the challenges to designing effective systems of private governance are even more complex than for ICCAs. How do we ensure accountability for management by private owners, and how can permanence of the protected status of that property be guaranteed for the long term? Can these areas even be considered as protected areas? These are legal and policy questions, but the ultimate success of any approach relying on alternative forms of governance to enhance protected area networks will depend on a scientifically robust monitoring and indicator framework that can demonstrate the effectiveness of alternative approaches.

Private lands have an especially important role as countries put in place mainstreaming approaches that include mitigation and compensation requirements for major production sectors, such as energy, mining, and infrastructure. There are now 27 state programs globally that require implementation of the mitigation hierarchy (avoid impacts, minimize impacts, offset/compensate for residual impacts) implying a target of no net loss of biodiversity of natural habitat as a result of these development activities. Many of these either allow for or explicitly call for investment in new protected areas as a possible compensation mechanism for unmitigated impacts. For example, the government of Colombia in 2012 approved a regulation requiring mandatory offsets for all projects subject to environmental licensing in the country. The license applicant will have to develop and maintain government-determined priority areas for conservation that are intended to compensate for the inevitable losses from development. These offset sites are explicitly planned to contribute to the creation and consolidation of areas in the National System of Protected Areas (SINAP) (Ministry of Environment and Sustainable Development, 2012). Though not without controversy, biodiversity offset mechanisms can be an efficient tool for increasing the connectivity and extent of land under conservation management and, ultimately, balancing conservation and development (McKenney & Kiesecker, 2009; BBOP, 2012).

Scaling up ambition and achieving sustained public support

The world's governments have formally agreed to a set of global protected area targets that appear to be realistic and attainable. The record on national-level commitments and implementation thus far indicates both action and intent to follow through on these commitments. However, this commendable record falls short on at least two fronts: The momentum generated thus far may not be self-sustaining in an era of increasing pressures on the land and waters and declining state budgets; and more importantly, the targets specified may be insufficient to halt biodiversity loss even if fully implemented (Brooks et al., 2004; Rodrigues et al., 2004). Keeping the momentum for action going and spurring the still deeper commitments that many believe are necessary will paradoxically require that protected areas not be viewed in isolation but as a vehicle to achieve a broad range of conservation and other societal goals.

Scaling up the targets themselves should not just be a numbers game; it must have a purpose. It must include efforts such as improving the functionality of protected areas at a landscape level or increasing connectivity among sites. Protected areas can be a cost-effective means of achieving many other Aichi Targets, including habitat loss (Target 5), anthropogenic pressures on coral reefs (Target 10), species status (Target 12), genetic diversity of crops (Target 13), restoration of ecosystem services (Target 14), and climate resilience (Target 15). To strengthen support for protected areas, policy makers should not place them in a box as an isolated strategy but explicitly integrate them as an efficient and cost-effective method for achieving a broad set of conservation objectives.

Linkages to the climate change agenda can also support expanding protected area networks. Protected areas can be an important climate change response strategy by helping buffer local climate variability; reducing the impacts of droughts, storms, and flooding; and maintaining ecosystem integrity and ecosystem services. The reality of climate change further calls for increasing the size and connectivity of protected areas to provide opportunities for species mobility (Dudley et al., 2010).

Perhaps the most significant qualifier for Aichi Target 11 is that protected areas should cover zones important for their ecosystem services. Ecosystem services define nature in relation to the benefits that people derive from it, such as water, food, storm protection, and cultural values. Ecosystem services are becoming an increasingly important framing device for discussing conservation and the value of nature (MEA, 2005; CBD, 2010c; TEEB, 2010). The use of protected areas to provide coastal protection against storms and sea-level rise and their role in providing food security (particularly fish) or for carbon storage provides a new way of thinking about protected areas as natural infrastructure and a viable alternative to building infrastructure. It also provides potential means to build new constituencies for protected areas based on the communities that benefit from those services.

While protected areas clearly support many environmental treaties and targets, more significantly, they need to be seen as a core *development* strategy. After the Rio +20 Summit in 2012, governments embarked on the negotiation of the Sustainable Development Goals (SDGs), adopted by the UN General Assembly in late 2015. Unlike the earlier Millennium Development Goals, the SDGs are intended to apply to developed and developing countries alike, providing a universal foundation for action. Protected areas are an obvious tool for achieving SDG 14 "Conserve and sustainably use the oceans, seas and marine resources" and SDG 15 "Protect, restore and promote sustainable use of terrestrial ecosystems", but they can also contribute to a number of other goals and targets:

Sustainable Development Goal	Protected areas contribution
Goal 1. End Poverty	Indigenous and community-managed protected areas support targets 1.4 and 1.5
Goal 2. End hunger, achieve food security and improved nutrition and promote sustainable agriculture	PAs support target 2.4, on maintenance of ecosystems, and 2.5, on plant genetic diversity
Goal 3. Ensure healthy lives and promote well-being	PAs provide recreational benefits and support mental health and wellbeing (target 3.4)
Goal 6. Ensure availability and sustainable management of water and sanitation	PAs can protect watersheds and water-related ecosystems (target 6.6) and support integrated water resource management
Goal 11. Make cities and human settlements inclusive, safe, resilient and sustainable	PA contribute to targets 11.4 on the safeguarding natural heritage, 11.5 on disaster risk reduction, 11.7 public green space

Goal 12. Ensure sustainable consumption and production patterns	Protection supports target 12.2 on sustainable management of natural resources
Goal 13. Take urgent action to combat climate change and its impacts	Protected areas provide carbon sequestration and are a resilience/adaptation strategy

Protected areas can serve as mechanisms to help meet many of the SDGs, but this knowledge is not well understood by many outside of the conservation community. Conservation scientists must be prepared to engage with human development initiatives in a measured and sustained fashion and demonstrate the advantage of this approach to policy makers who may view it as a distraction at best and antagonistic to development goals at worst. Conservationists must reframe the costs versus benefits debate and move away from the notion that protected areas are a developed country construct and somehow inimical to local values and needs in other parts of the world. Developing the case for protection as a cross-cutting element in the SDGs will require data as well as reasoned argument; acquiring this hard evidence should be on the core agenda of the Intergovernmental Science-Policy Platform on Biodiversity and Ecosystem Services (IPBES).

As much as ecological precepts have largely defined the scope of protected areas to date, demonstrating the centrality of ecosystems to human health and prosperity will be the urgent task to retain and build on past success. The value of protected areas will be enhanced if their application is embedded across a web of interrelated policy agendas, goals, and obligations. Scaling up effectively means integrating protected areas into the development agenda and protected areas must not be seen as a residual strategy but rather as a core strategy for development.

Notes

1. The United States signed but has not ratified the CBD. It participates as an observer.
2. The 2010 overall target was "To achieve by 2010 a significant reduction of the current rate of biodiversity loss at the global, regional and national level as a contribution to poverty alleviation and to the benefit of all life on earth," adopted in 2002 at the COP 6 in the Hague.
3. The full text of Target 11 reads: By 2020, at least 17% of terrestrial and inland water and 10% of coastal and marine areas, especially areas of particular importance for biodiversity and ecosystem services, are conserved through effectively and equitably managed, ecologically representative, and well-connected systems of protected areas and other effective area-based conservation measures and integrated into the wider landscapes and seascapes.
4. Regional Fisheries Management Organisations (RFMOs) may also in some cases designate MPAs on the High Seas, but these are only binding on parties to the given RFMO; they cannot restrict third-party behavior.

References

Baettig, M. B., Brander, S., & Imboden, D. M., 2008. Measuring countries' cooperation within the international climate change regime. *Environmental Science and Policy*, 11(6), p. 478–489.

BBOP, 2012. *Standard on biodiversity offsets*, Washington, DC: BBOP.

Bertzky, B. et al., 2012. *Protected planet report 2012: Tracking progress towards global targets for protected areas*, Gland & Cambridge: IUCN & UNEP-WCMC.

Brooks, T. M. et al., 2004. Coverage provided by the global protected area system: Is it enough. *BioScience*, 54(12), p. 1081–1091.

CBD, 2004. *Decisions adopted by the CoP to the CBD at its 7th meeting (UNEP/CBD/COP/7/21)*, Montreal: Secretariat of the CBD.

CBD, 2010a. *Decisions adopted by the CoP to the CBD at its 10th meeting (UNEP/CBD/COP/DEC/X/2)*, Montreal: Secretariat of the CBD.

CBD, 2010b. *Global biodiversity outlook 3*, Montreal: Secretariat of the CBD.

CBD, 2010c. *Recommendations adopted by the WGRI at its 3rd meeting (3/5)*, Montreal: Secretariat of the CBD.

CBD, 2012. *Protected areas: Progress in the implementation of the programme of work and achievement of Aichi biodiversity target 11 (UNEP/CBD/COP/11/26*, Montreal: Secretariat of the CBD.

Chape, S., Harrison, J., Spalding, M., & Lysenko, I., 2005. Measuring the extent and effectiveness of protected areas as an indicator for meeting global biodiversity targets. *Philosophical Transactions of the Royal Society of London Series B*, 360(1454), p. 443–455.

Chayes, A. & Chayes, A. H., 1993. On compliance. *International Organization*, 47(2), p. 175–205.

Dillon, B., 2004. The use of the categories in national and international legislation and policy. *PARKS*, 14(3), p. 15–22.

Downs, G. W., Rocke, D. M. & Barsoom, P. N., 1996. Is the good news about compliance good news about cooperation. *International Organization*, 50(3), p. 379–406.

Dudley, N. et al., 2010. *Natural solutions: Protected areas helping people cope with climate change*, Gland/Washington, DC & New York: IUCN-WCPA, TNC, UNDP, WCS, World Bank & WWF.

Dyer, M. I. & Holland, M. M., 1988. UNESCO's man and the biosphere programme. *BioScience*, 38(9), p. 635–641.

Ervin, J. & Gidda, S., 2012. *Input to the report of the high level panel on global assessment of resources for implementing the strategic plan for biodiversity 2011–2020: Target 11*, Hyderabad: UNDP & CBD.

Leverington, F. et al., 2010. A global analysis of protected area management effectiveness. *Environmental Management*, 46(5), p. 685–698.

Ministry of Environment and Sustainable Development, 2012. *Manual for allocating offsets for loss of biodiversity*, Bogota, Republic of Colombia.

McKenney, B. A. & Kiesecker, J. M., 2009. Policy development for biodiversity offsets: A review of offset frameworks. *Environmental Management*, 45(1), p. 165–176.

MEA, 2005. *Ecosystems and human well being: Biodiversity synthesis*, Washington, DC: World Resources Institute.

Mitchell, R. B., 2011. Transparency for governance: The mechanisms and effectiveness of disclosure based and education based transparency policies. *Ecological Economics*, 70(11), p. 1882–1890.

Phillips, A., 2004. The history of the international system of protected area management categories. *PARKS,* 14(3), p. 4–14.

Rodrigues, A. S. L. et al., 2004. Effectiveness of the global protected area network in representing species diversity. *Nature*, 428, p. 640–643.

Spalding, M. D. et al., 2013. Protecting marine spaces: Global targets and changing approaches. *Ocean Yearbook*, 27, p. 213–248.

TEEB, 2010. *The economics of ecosystems and biodiversity: Mainstreaming the economics of nature—A synthesis of the approach, conclusions and recommendations of TEEB*, Malta: Progress Press.

UNESCO, 2013. Mongolia Sacred Mountains: Bogd Khan, Burkhan Khaldun Khaldun, Otgon Tenger. [Online] Available at: http://whc.unesco.org/en/tentativelists/936/ [Accessed March 26, 2013].

United Nations, 2012. *Resolution adopted by the general assembly: 66/228 The future we want*, New York: United Nations.

WDPA, 2010. *The World Database on Protected Areas (WDPA)*, Gland & Cambridge: IUCN, UNEP-WCMC.

2

Protected Area Diversity and Potential for Improvement

N. Dudley and S. Stolton

Equilibrium Research, Bristol, UK

Introduction

What is the common link amongst a disused gravel pit in the British East Midlands, a Switzerland-sized wilderness of limestone karst in northern Canada, a rainforest managed by indigenous peoples for medicinal herbs in Colombia and a sacred mountain visited by over eight million people a year outside Seoul, Korea? These and tens of thousands of other land and water sites around the world are all protected areas, set aside by governments, communities and private individuals to help maintain global biodiversity. Yet protected areas are often treated as if they conform to a single, narrow management approach by conservation biologists and social scientists. The IUCN definition of a protected area embraces a surprisingly wide variety of both management models and governance types. This provides multiple conservation opportunities, some of which we are only now starting to explore. A better understanding of the range of options could lead to more informed choices in national and regional conservation strategies. In this chapter, we explore ways to increase the potential of protected areas by:

- Describing the full variety of protected area options
- Giving examples of how countries are using different approaches to build and maintain their protected area systems
- Identifying some gaps in our knowledge and necessary next steps

Protected Areas: Are They Safeguarding Biodiversity?, First Edition.
Edited by Lucas N. Joppa, Jonathan E. M. Baillie and John G. Robinson.

The variety of protected areas

IUCN has spent several decades wrestling with the question of what defines a protected area and how and what protected areas contribute to human society. The latest manifestation of this thinking, following a 4-year consultation period that itself built on a lengthy research project (Bishop et al., 2004), was an agreement of a new definition of a protected area and publication of revised guidelines to the IUCN protected area management categories and governance types (Dudley, 2008). The new definition of a protected area is 'a clearly defined geographical space, recognised, dedicated and managed, through legal or other effective means, to achieve the long term conservation of nature with associated ecosystem services and cultural values'.

This replaces the 1994 definition, 'an area of land and/or sea especially dedicated to the protection and maintenance of biological diversity, and of natural and associated cultural resources, and managed through legal or other effective means' (IUCN, 1994). The new definition ends a decades-long disagreement amongst IUCN members by stating clearly that nature conservation is the primary function of protected areas (which is further emphasised by an associated principle), although it also stresses the importance of ecosystem services and cultural benefits, which themselves require management attention (Dudley et al., 2010a). Other key points are that the definition embraces more than just state-run, legally established protected areas, includes the need for management effectiveness (which was absent in earlier iterations) and broadens objectives from biological diversity to nature conservation, thus embracing geological diversity and important landforms.

Protected areas that meet the IUCN definition can be subdivided in many different ways: by habitat type, ecoregion, biome, management effectiveness and so on. There have also been suggestions to consider a classification based on biodiversity targets (Boitani et al., 2008). IUCN currently works with two typologies, based around management categories and governance types, as outlined in Box 2.1. Any protected area category can be applied to any governance type, as shown in the protected area matrix in Table 2.1. All protected area management approaches are recognised by IUCN as being useful to conservation, but they are not interchangeable in any simplistic way, nor are they equally useful in every situation. Once again, this marks a change in 2008, before then IUCN stated that all protected area categories were equal without elaborating that equality is not necessarily the same as equivalence in every situation.

Together, the management category and governance type provide a clear picture of what the protected area is aimed to achieve and how and who has the authority to set management objectives.

The effectiveness of the various approaches is still the subject of debate. Some proponents have been critical of the less strictly protected approaches (e.g. Locke & Dearden, 2005), while conversely, others believe the traditional, exclusionary

Box 2.1 **Typologies of protected areas**

IUCN currently subdivides protected areas according to both management approach and governance type. Six different **management categories** of protected areas are defined (one with a subdivision):

Ia – Strict nature reserve: mainly for science or wilderness protection
Ib – Wilderness area: mainly for wilderness protection
II – National park: mainly for ecosystem protection and recreation
III – Natural monument: mainly for conservation of specific natural features
IV – Habitat/species management area: mainly for conservation of species and habitat
V – Protected landscape/seascape: mainly for landscape/seascape conservation or recreation
VI – Managed resource protected area: mainly for the sustainable use of natural resources

In addition, a series of four main governance types are recognised, each with some subdivisions. Governance refers to who makes decisions about management within the protected area:

A – By governments: national or local government, or by bodies appointed by a government
B – Through co-management: ranging from local partnerships to trans-boundary protected areas
C – By private bodies: individuals, not-for-profit trusts or for-profit private ecotourism reserves
D – By indigenous peoples and local communities: anything from local village-run reserves to large indigenous peoples' territories

protected area model is ineffective and socially inequitable (Dowie, 2009; Fairhead et al., 2012). Several of the other chapters in this book report on research currently underway to address some of these questions. One important interim result of the various criticisms is that the nature of protected area management is changing in many countries. Some previously strict reserves with no human access are opening their borders to greater use by local communities, while some of the looser protected area designations are being tightened to increase their effectiveness for conservation. In other words, management strategies are often still evolving as we learn more about what does and doesn't work.

Table 2.1 **'The IUCN protected area matrix': a classification system for protected areas comprising both management category and governance type**

Governance types	**A. Governance by government**			**B. Shared governance**			**C. Private governance**			**D. Governance by indigenous peoples and local communities**	
Protected area categories	**Federal or national ministry or agency in charge**	**Subnational ministry or agency in charge**	**Government-delegated management (e.g. to an NGO)**	**Transboundary management**	**Collaborative management (various forms of pluralist influence)**	**Joint management (pluralist management board)**	**Declared and run by individual landowner**	**...by non-profit organisations (e.g. NGOs, universities, cooperatives)**	**...by for-profit organisations (e.g. individual or corporate landowners)**	**Indigenous peoples' conserved areas and territories – established and run by indigenous peoples**	**Community conserved areas – declared and run by local communities**
Ia. Strict Nature Reserve											
Ib. Wilderness Area											
II. National Park											
III. Natural Monument											
IV. Habitat/Species Management											
V. Protected Landscape/ Seascape											
VI. Managed Resource Protected Area											

Source: From Dudley (2008).

Different approaches to building national protected area systems

What does this mean in practice? The history and politics of land use, geography, geology, the age of the protected area network and related factors such as the extent to which local communities are organised, relative wealth of a country and public attitudes to conservation together combine to shape national protected area systems. Few, if any, match theoretical models described in conservation textbooks, neither is their location determined solely by conservation considerations. Conservation everywhere tends to be one of the weaker parts of government. Partly in consequence, conservation models necessarily have to be pragmatic in acknowledging the limits of what is possible and therefore usually imperfect in delivery. In the following section, some examples are given to illustrate the wide variety of approaches already in existence.

United Kingdom: Large areas but low protection

On paper the United Kingdom has, relative to its size, one of the most extensive protected area systems in the world. But the overwhelming majority of the area involved is in the form of IUCN Category V protected landscapes, both national parks and other designations such as Areas of Outstanding Natural Beauty, with rather weak biodiversity conservation legislation. There are also many small protected areas in private hands, currently well over 3500 not counting those of the National Trust, which is the largest landowner in the United Kingdom. Most of the protected landscapes were set up at a time when issues of landscape aesthetics and access had a much higher political importance than biodiversity conservation and attempts to retrofit stricter conservation have met with variable success. There are now emerging attempts to strengthen conservation management within protected landscapes and a debate about whether all designated sites are actually protected areas as recognised by IUCN (e.g. Crofts et al., 2014).

Finland: Northern wilderness with indigenous use

Similarly, Finland has large state protected areas although unlike the United Kingdom, many of these are much more strictly protected, in Category Ib or II. However, here, the geography is skewed; there is a low level of protection in the south and at a relative scale very large protected areas in the north, where wilderness areas have been set up through agreements with the indigenous Sami people. The Sami lay claim to much of the territory within the Arctic Circle, and protection involves a trade-off between indigenous peoples and the state that has resulted in the establishment of large protected areas but with the caveat that these still permit both reindeer grazing by the Sami and traditional hunting within their borders. Most studies suggest that many northern Finnish protected areas are currently overgrazed, resulting in significant changes to ground vegetation (Gilligan et al., 2005). More recently, many small nature reserves have been established, often by private landowners who no longer manage their woodlands for anything except recreation and environmental values.

Tanzania: Massive, state-run protected areas and private reserves

Almost a quarter of the country is subdivided into a series of very large state-run national parks, mainly Category II, where local use is banned or strictly controlled. One exception is Ngorongoro crater, where some degree of traditional grazing is still allowed and management structures are different. Tourism associated with national parks is the single largest source of foreign income, and tourism access can sometimes heavily influence management decisions. But many of these areas also experience severe poaching off take (Holmern et al., 2004) and human settlements have increased around the edges of some national parks in consequence (Wittemyer et al., 2008). In addition, large privately owned hunting areas are often included within consideration of the protected area estate; although set up for shooting game, populations remain stable and these private sites may provide more effective protection than state-run protected areas (Stolton et al., 2014).

Madagascar: Rethinking protected landscapes in the face of land shortages

A declaration by the then president of Madagascar to triple the coverage of protected areas in 2003 created a quandary for Madagascar conservation organisations: how to increase coverage in a densely populated, desperately poor but also highly diverse country with almost unique levels of endemism. Several approaches were taken, and the situation was further confused by a subsequent coup and associated increase in illegal logging and poaching within existing protected areas. One particular aspect of relevance here is the emergence of a Malagasy-specific definition of Category V, developed through extensive consultation and fitted to conditions within the country (Borrini-Feyerabend & Dudley, 2005). Whereas most countries interpret protected landscapes as being generally large areas with fairly loose definition and multiple ownerships, Madagascar developed an approach that focused much more on quite small, culturally and often spiritually important sites with high conservation value. The significance of such approaches beyond Madagascar has yet to be considered.

South Korea: Moving towards stricter protection

From being one of the most deforested countries in the world in the 1970s, South Korea undertook a massive forest restoration programme and also developed an extensive network of protected areas. Protected areas are strongly supported by the general public in Korea, with annual visitor numbers reaching a staggering eight million a year for the national park nearest to Seoul, although reactions from communities living inside or near national parks are more mixed. Coverage is biased towards mountains and forests, with very low levels of protection in lowland or culturally managed areas and poor protection for coastal areas, where habitat loss remains a subject of international controversy (KNPS, 2009). Against the trends of recent moves to liberalise controls on the use of protected areas, the government of South Korea has worked with IUCN to re-categorise several of its protected areas from V to II, signalling a stricter protection regime following changes in legislation, depopulation from rural areas and some revisions of boundaries to focus management on areas with high nature conservation value (Shadie et al., 2012).

India: State-run protected areas and community conservation

Two systems run in parallel. There is an extensive and centrally controlled system of protected areas, including some of the oldest national parks in the world established under British rule to protect species such as the rhino. These have an exclusionary approach with no permission for human populations living inside the park boundaries and no use for subsistence; such controls are not always applied in practice. But alongside the formal national park network is a separate and largely unrecognised network of community conserved areas and sacred natural sites, usually far smaller and maintained by local communities, sometimes over centuries (Pathak, 2009). Recent attempts to quantify these have estimated, for instance, that there are between 100,000 and 150,000 sacred groves in 19 out of 28 states of India (Subash Chandran & Hughes, 2000). These two approaches still largely run independently of each other.

Kuwait: Protection for restoration

The country has been severely impacted by overgrazing, off-road vehicles, poaching and the catastrophic impacts of the 1991 war, with the associated burning of over 500 oil wells, massive oil spillage and deposition of a thick deposit known as 'tarcrete' over large parts of the country. A network of protected areas focuses on natural dry land vegetation, some offshore islands and coastal wetlands. Several of the existing and proposed protected areas already require extensive restoration. Unusually, most of the larger protected areas are fenced, in part for security reasons (many lie close to the border with Iraq). While fencing prevents the overgrazing prevalent in the rest of the country, a total lack of grazing is also likely to prove detrimental in the long term.

Colombia: A combination of private, community and state approaches

Like several countries in Latin America, Colombia combines a large state-run protected area system, predominantly categorised for fairly strict protection, with additional and extensive protected areas run by local authorities, private individuals, communities and indigenous peoples. State protected areas, which cover the largest area, are managed in different ways depending on circumstance; the latest one to be agreed, for instance, is co-managed with the local indigenous people. Security considerations affect management, on the one hand, protecting some areas from exploitation simply because they are too dangerous but conversely creating a general environment of lawlessness that encourages illegal use. As security conditions change, protected area management also has to adapt.

This sample list gives a flavour of the kinds of approaches and the sorts of developments; few, if any, countries have identical protected area philosophies, and multiple philosophies can exist within a single nation state, sometimes diametrically opposed. As countries understand more about how to manage their protected areas, as economies and social aspirations change and as practitioners have chances to share ideas and experience, we may see approaches coalescing around a few successful models or conversely might see an even greater proliferation of approaches. Critical to progress is an understanding both what we know and what we do not know. In the third part below, some important questions are raised that might help guide future research priorities.

Identifying some gaps in our knowledge and necessary next steps

Much of this book is concerned with reporting on research that considers whether or not protected areas deliver their intended conservation benefits. Surprisingly, until quite recently, this question has not been addressed in any systematic way and great progress has been made in the past 5 years and we can now say, with some confidence, that at a general level most protected areas are effective at conserving most biodiversity. However, we remain at the beginning of any more sophisticated understanding, with a lot of work still to do. Below we list seven research questions that require urgent consideration. Six of them address the biodiversity considerations raised in this volume, particularly issues relating to size and connectivity; the seventh considers how these ecological issues can be situated in a wider strategic context. They are all important, but the list does not attempt to be exhaustive.

Question 1: What is happening in Category V?

Protected landscapes have increased more rapidly than other protected area management categories in the last few years, for instance, covering over half of the protected area estate by area in Europe (Gambino, 2008). It is not hard to see why. Category V embraces cultural landscapes, living communities and activities such as agriculture and forestry, making it far more agreeable to politicians and rural communities than stricter and more exclusive protected areas. But does it work? We have already referred to the debate stirred by Locke & Dearden (2005), which elicited some thoughtful responses (Mallarach et al., 2008). More recently, we compiled case studies of wild biodiversity conservation in protected landscapes (Dudley & Stolten, 2012), which provide some encouraging signs of success but still woefully few hard data. Yet many protected landscapes exist in countries where biodiversity monitoring is organised and capable of generating quantitative trend information over time. Developing well-designed and carefully implemented studies on the effectiveness of Category V protected areas and, critically, the factors affecting their success or failure needs to be an important priority for those interested in protected area management effectiveness and conservation success.

Question 2: Under what conditions do networks of small reserves interact ecologically?

The theory of island biogeography tells us that small reserves will tend to do less well than large reserves and the 'single large' or 'several small' debate has rumbled for decades. Many factors are relevant, but from a connectivity perspective, we still know worryingly little about what constitutes an island. This will vary depending on individual species, the environment found between patches of habitat, the time of year and seasonal climatic conditions (see Figure 2.1). As conservationists are forced more and more frequently to work in fragmented landscapes and seascapes, it becomes increasingly important to know when these constitute a series of isolated and degrading patches and when they might interact together to form something more akin to a mega-reserve (see Figure 2.2).

Although some efforts have been made to quantify these links, for instance, for UK woodland habitats (Peterken, 2002) and under various logging regimes (Lindenmayer et al., 2015), our knowledge is in most cases still extremely partial, often based on isolated observations and wholly inadequate for the practical needs of management. A thousand doctoral research projects could help.

Question 3: What do small strict reserves contribute to surrounding, less strictly protected areas?

This question links to both the previous issues. Most large protected landscapes, at least in Europe, contain smaller, more strictly protected reserves within their

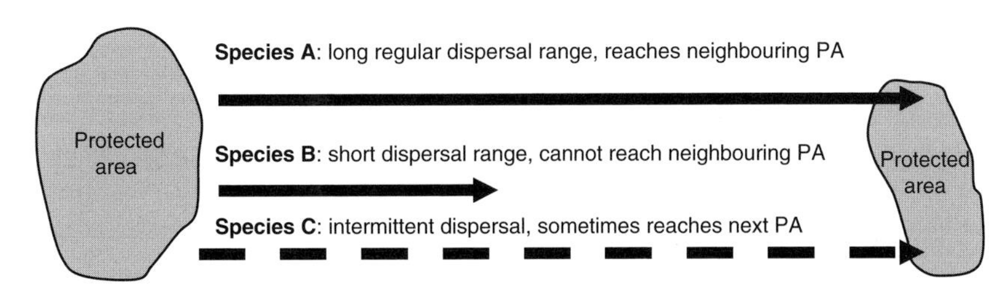

Figure 2.1 **Dispersal between protected areas will vary with species, intervening environment and many other factors**

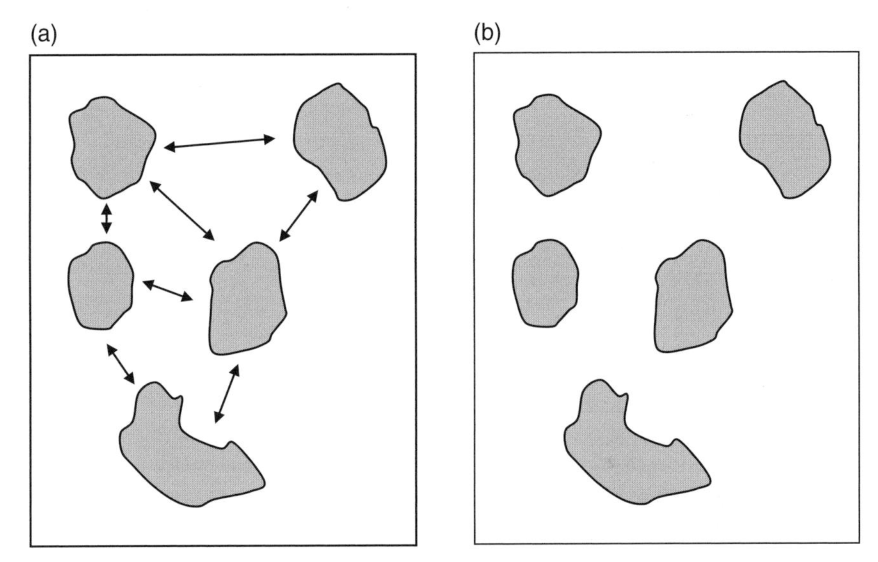

Figure 2.2 **Different options for connectivity between protected areas: (a) network of small reserves functioning as a mega-reserve and (b) network of small reserves isolated from each other and not interacting**

borders. Even when we have some quantitative data on the success or failure of protected landscapes, studies do not generally distinguish the extent to which this is due to the various management regimes and nested protected areas within the protected landscape or to the whole landscape in general. We understand little about when and if there is spillover from reserves into the wider environment, with the exception of marine protected areas where there is good evidence that such spillover occurs (e.g. Halpern, 2003). In data-poor countries, we do not in many cases even know how particular species relate to the protected area network, that is, whether they are virtually confined to protected areas, have a distribution unaffected by the presence of protected areas or some other relationship (see Figure 2.3). Studies within protected areas only give limited information

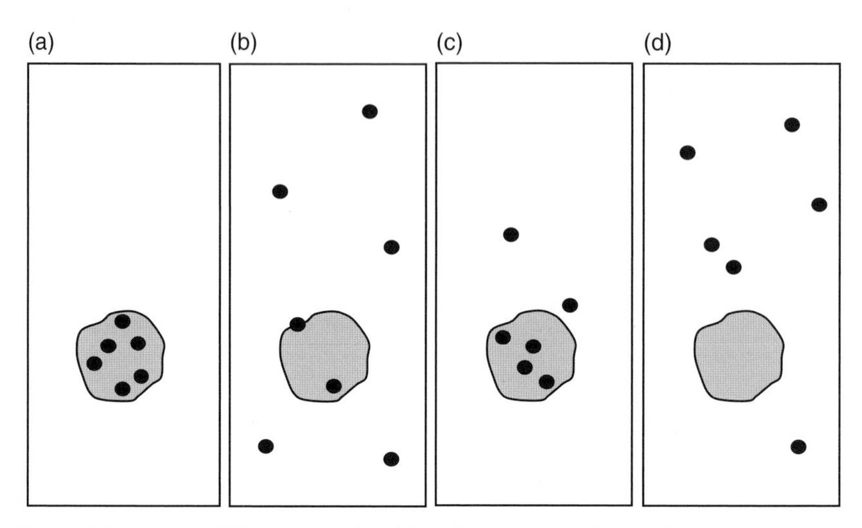

Figure 2.3 **Some different relationships between species and protected areas: (a) species confined entirely inside protected areas, (b) species distribution unaffected by protected areas, (c) species concentrated in but not confined to protected areas, and (d) species entirely outside protected areas**

and more widespread landscape-scale studies are required (which in some cases might include reanalysis of existing data).

Question 4: When does good management in small reserves 'out compete' bad management in large reserves?

Preliminary research reported elsewhere in this volume suggests that small reserves are surprisingly effective at conserving biodiversity, despite the inherent limitations of their size. It is important that we understand why. Alongside issues of connectivity, it may be that small reserves receive proportionately higher levels of management than larger protected areas, and that management effort is partly or wholly counteracting size limitations. However, if this is true it has important implications in terms of long-term management efforts, with attendant costs. Currently, many small reserves are either managed by volunteers in the more developed countries or by communities who themselves need the ecosystem services or resources provided by the protected area in poorer countries. We cannot predict with any confidence whether either of these conditions will continue in the long term or the implications for conservation strategies if they do not. A good start would be to know more about the role of management interventions in maintaining biodiversity in small reserves, options for increasing connectivity and whether management can offset the natural disadvantages of small reserves in the long term.

Question 5: How do different governance types compare in terms of effectiveness?

To date, the large majority of studies on management effectiveness for biodiversity conservation have focused on state-run protected areas. There are a few exceptions: Madhu Rao et al. (this volume, Chapter 10) report on a meta-study of indigenous and community conserved areas in this volume, conservation within sacred natural sites has been carefully studied in several countries (Dudley et al., 2010b), and private protected areas have started to receive attention (Fisher & Dills, 2012). But these preliminary studies raise as many questions as they supply answers. When can there be a mismatch between community perceptions of ecosystem health and the reality? Does community conservation tend to favour particular groups of species at the expense of others? Are community and private reserves generally smaller and if so does this impact on their effectiveness? We are still at the beginning of our understanding of these issues.

Question 6: How do we implement a landscape approach?

All the preceding questions contribute to one larger question, how do we fit a collection of disparate strategies together into one coherent and workable system? Landscape approaches have been promoted for over a decade now, but there is still remarkably little clear guidance about what this means in practice, with the fall-back position often being simply to ask for more protected areas. There has been progress in understanding connectivity conservation (Worboys et al., 2010), but this is only one piece of a wider puzzle. Conservation plans aimed at single species and groups of species, such as the Biodiversity Action Plans promoted in the European Union, can undermine broader conservation objectives by focusing on a narrow target without sufficient consideration of the consequences for other species or wider ecosystem health. The need for proper 'joined-up thinking' is dramatically increased by environmental instability as a result of both immediate development pressures and more intransigent climate change. Yet currently, we do not even have a coherent landscape approach framework to help people start. Most conservation planning remains at the stage of designing protected area systems without consideration of how they fit into wider landscape plans; the few exceptions, such as the CAPE Action Plan in South Africa, show what is possible but remain isolated from overall guidance. A multidisciplinary effort to build understanding and practical experience of landscape approaches would need to incorporate all the earlier questions, along with consideration of many other variables. Working out a way to approach this that is not impossibly complicated and expensive is perhaps the single largest conservation challenge for the second and third decades of the 21st century.

Question 7: How do we integrate other values into our protected area planning?

Amongst the wider issues to be addressed is the relationship between conservation and other land and water uses. Experience shows that it is difficult to dedicate the existing protected area estate to conservation, let alone an expanded protected area system or the other elements in a landscape mosaic. To recall, the protected area definition identifies the importance of associated ecosystem services and cultural values but few conservation planning processes integrate these in anything but a simplistic or ad hoc manner. Yet protection of such services often (usually) results in good biodiversity conservation as well, truly a win-win. Understanding what other benefits a protected area can supply and communicating these to the beneficiaries can strengthen the political situation of a protected area, generate additional funds and build local support. Yet despite the impact of high-profile projects such as The Economic or Ecosystems and Biodiversity (TEEB) and isolated studies in individual protected areas (van Beukering et al., 2003), there is little guidance for practitioners. Next steps need to include faster ways of working out the range and value of such benefits (e.g. the InVEST methodology or Dudley and Stolton, 2009) and ways of incorporating such values into planning at a landscape scale without undermining overarching biodiversity conservation objectives. In a way, Question 7 is an extension of Question 6, incorporating wider social and economic values into what otherwise often has a narrow ecological remit. This in turn implies that there will be trade-offs between conservation and other aims, particularly outside protected areas.

Protected areas have lost intellectual ground in the last decade and are regarded by at least some of the conservation and social development communities as outmoded and inefficient conservation tools, which carry a high social and economic price. The results reported in this volume provide convincing evidence that this pessimism is misplaced. It is now up to us to fill in some of the remaining gaps in our understanding and ensure that a comprehensive and well-managed protected area system becomes a respected, popular and politically essential component of national development plans.

References

van Beukering, P. J. H., Cesar, H. S. J. & Janssen, M. A., 2003. Economic valuation of the Leuser National Park on Sumatra, Indonesia. *Ecological Economics*, Volume 44, p. 43–62.

Bishop, K., Dudley, N., Phillips, A. & Stolton, S., 2004. *Speaking a common language: The uses and performance of the IUCN system of management categories for protected areas*, Cardiff & Gland: Cardiff University, IUCN & UNEP WCMC.

Boitani, L. et al., 2008. Change the IUCN protected area categories to reflect biodiversity outcomes. *PLoS Biology*, Volume 6, p. 436–438.

Borrini-Feyerabend, G. & Dudley, N., 2005. *Les aires protegees a Madagascar: Batir le systeme a partir de la base*, Gland: IUCN WCPA & IUCN CEESP.

Dowie, M., 2009. *Environmental refugees*, London: Earthscan.

Dudley, N., 2008. *Guidelines for applying protected area management categories*, Gland: IUCN.

Dudley, N. & Stolton, S., 2009. *The Protected Area Benefits Assessment Tool*, Gland: WWF International.

Dudley, N. & Stolton, S., 2012. *Protected landscapes and wild biodiversity*. Volume 3 in the values of protected landscapes and seascapes series, Gland: IUCN, p. 7–18.

Dudley, N. et al., 2010a. Conservation of biology in sacred natural sites in Asia and Africa: A review of the scientific literature. In: B. Verschuuren, R. Wild, J. McNeely & G. Oviedo, eds. *Sacred natural sites: Conserving nature and culture*, London: Earthscan, p. 19–32.

Dudley, N., Parrish, J. D., Redford, K. H. & Stolton, S., 2010b. The revised IUCN protected area management categories: The debate and ways forward. *Oryx*, Volume 44, p. 485–490.

Fairhead, J., Leach, M. & Scoones, I., 2012. Green grabbing: A new appropriation of nature. *The Journal of Peasant Studies*, Volume 39, p. 237–262.

Fisher, J. R. B. & Dills, B., 2012. Do private conservation activities match science based conservation priorities. *PLoS ONE*, Volume 7, p. e46429.

Gambino, R., 2008. *Parchi d'Europa. Verso una politica europea per la aree protette*, Pisa: ETS Edizioni.

Gilligan, B., Dudley, N., Fernandez de Tejada, A. & Toivonen, H., 2005. *Management effectiveness evaluation of Finland's protected areas*, Helsinki: Metsahallitus.

Halpern, B. S., 2003. The impact of marine reserves: Do reserves work and does reserve size matter. *Ecological Applications*, Volume 13, p. 117–137.

Holmern, T. et al., 2004. *Human-wildlife conflicts and hunting in the western Serengeti, Tanzania*, Trondheim: NINA.

IUCN, 1994. *Guidelines for protected area management categories*, Gland & Cambridge: IUCN.

KNPS, 2009. *Korea's protected areas: Evaluating the effectiveness of South Korea's protected area system*, Gland, Cambridge & Island of Jeju: IUCN & KNPS.

Lindenmayer, D. B., Wood, J., McBurney, L. Blair, D. & Banks, S. C., 2015. Single large versus several small: The SLOSS debate in the context of bird responses to a variable retention logging experiment. *Forest Ecology and Management*, Volume 339, p. 1–10.

Locke, H. & Dearden, P., 2005. Rethinking protected area categories and the new paradigm. *Environmental Conservation*, Volume 32, p. 1–10.

Mallarach, J. M. et al., 2008. In defence of protected landscapes: A reply to some criticisms of category V protected areas and suggestions for improvement. In: N. Dudley & S. Stolten, eds. *Defining protected areas: An international conference in Almeria, Spain*, Gland: IUCN, p. 31–38.

Pathak, N., 2009. *Community conserved areas in India: A directory*, Pune: Kalpavriksh.

Peterken, G. F., 2002. *Reversing the habitat fragmentation of British woodlands*, Godalming: WWF.

Roger Crofts, Nigel Dudley, Chris Mahon, Richard Partington, Adrian Phillips, Stewart Pritchard, Sue Stolton (2014). *Putting Nature on the Map: A Report and Recommendations on the Use of the IUCN System of Protected Area Categorisation in the UK*. United Kingdom: IUCN National Committee UK.

Shadie, P., Heo, H. Y., Stolton, S. & Dudley, N., 2012. *Protected area management categories and Korea: Experience to date and future directions*, Gland & Seoul: IUCN & KNPS.

Stolton, S., Redford, K. R. H. & Dudley, N., 2014. *The futures of privately protected areas*, Gland: IUCN.

Subash Chandran, M. D. & Hughes, J. D., 2000. Sacred groves and conservation: The comparative history of traditional reserves in the Mediterranean area and in south India. *Environment and History*, Volume 6, p. 169–186.

Wittemyer, G. et al., 2008. Accelerated human population growth at protected area edges. *Science*, Volume 321, p. 123–126.

Worboys, G. L., Francis, W. L. & Lockwood, M., 2010. *Connectivity conservation management: A global guide*, London: Earthscan.

3

Sound Investments: Protected Areas as Natural Solutions to Climate Change and Biodiversity Conservation

K. MacKinnon

IUCN World Commission on Protected Areas (WCPA) Cambridge, UK

Introduction

Protected areas are widely recognized as one of the most effective ways to conserve biodiversity in both terrestrial and marine habitats (Bruner et al., 2001; Gaines et al., 2010; Selig & Bruno, 2010). Accordingly, in 2004, the parties to the Convention on Biological Diversity (CBD) adopted a Programme of Work on Protected Areas (PoWPA), which set targets for expanded coverage for both terrestrial and marine protected areas, improvement of management effectiveness, and recommendations on improving governance, management capacity, and sustainable finance. Although many of these targets still need to be realized, an increase in protected area coverage was one of the few positive biodiversity indicators reported in 2010. Subsequently, at the 10th Conference of the Parties (COP) to the CBD, the signatory countries adopted a new Strategic Plan with 20 headline targets to improve biodiversity conservation by 2020. Target 11 specifically identifies the need to expand global protected area coverage to at least 17% of the terrestrial and inland water habitats and 10% of the marine areas. Achieving Target 11 will also contribute significantly to the other 19 Aichi Targets.

There are now 200,000 protected areas covering 15.4% of the world's land area and 8.4% of coastal waters (Juffe-Bignoli, D. 2014). These facts do not tell the whole story. Although some countries have already protected 17% of their land area, many habitats and species are still poorly represented or not recorded within the global protected area network (Butchart et al., 2012). At present, only one third of the 823 terrestrial ecoregions meet the 17% target, while one

Protected Areas: Are They Safeguarding Biodiversity?, First Edition.
Edited by Lucas N. Joppa, Jonathan E. M. Baillie and John G. Robinson.

tenth still have less than 1% of their area protected. Similarly, by 2010, only 30 of the 232 coastal marine ecoregions met the 10% protection target, while 137 (59%) had less than 1% of their area protected (Bertzky et al., 2012; Woodley et al., 2012). Achieving Target 11 will therefore require a substantial increase in protected area coverage and some hard decisions on land and coastal use, especially in habitats that are currently poorly protected but have high potential for agricultural production.

In addition to more representative coverage, Target 11 prescribes other activities relevant to strengthening protected area management, including improved governance, connectivity, and better integration with surrounding land and seascapes. Establishing a fully representative and effectively managed system of protected areas will have significant costs (CBD, 2012; McCarthy et al., 2012). Generating the necessary political support and financing for such an expansion will require greater awareness and appreciation of the multiple social and economic benefits that protected areas can provide and much stronger linkages between the conservation agenda and development strategies (MacKinnon et al., 2011).

In the last decade, climate change has become the key environmental concern. Climate change will exacerbate the impacts of other threats and increase biodiversity loss. Indeed, the impacts of climate change are already being recorded in some natural ecosystems such as coral reefs, with 30% of the warm water corals in the Caribbean lost since the early 1980s (UNEP, 2006). Even a relatively modest 2°C global temperature rise will cause significant modifications to ecosystems and biodiversity loss (Bergengren et al., 2011; Bellard et al., 2012) but will also impact on human livelihoods and welfare. Food security, water stress, land degradation, and risks from natural disasters will be exacerbated by rising temperatures, more erratic rainfall, extreme climatic events, and rising pressure on land and natural resources. The impact will be especially hard in some of the poorest and least developed countries. Two thirds of the African continent, for instance, is desert or dry lands, and 75% of the agricultural dry land areas are already degraded. By 2020, 75–229 million people in Africa could be suffering severe water shortages, and some countries are expected to suffer a 50% reduction in rain-fed agriculture. Water shortages will impact on agricultural production, undermining efforts to reduce poverty and, in the worst cases, leading to famine and human migration. In Asia, coastal areas and especially densely populated delta regions will be subject to greater risk of flooding, while water shortages will impact on human health through the spread of waterborne diseases. Small island states are likely to suffer from sea-level rise, storm surges, impacts on coastal development, water shortages, and increased exposure to invasion by nonnative species (World Bank, 2010). These global impacts on human livelihoods, food security, human health, land available for agriculture, and coastal fisheries will have knock-on effects on the land and sea available for protected areas.

Protected areas as natural solutions to climate change

Climate change is thus a threat to biodiversity but also provides an opportunity to highlight the benefits and services that healthy ecosystems can provide in helping vulnerable communities to cope with environmental crises (World Bank, 2010; Dudley et al., 2010, 2013; Davies et al., 2012; MacKinnon, 2012). Protected areas are some of the most effective ways to maintain natural habitats and species; protect water supplies and other ecosystem services; and act as a safety net, helping communities to cope with and adapt to climate change, drought, and disaster.

There is already considerable evidence that protected areas can play an important role in storing and sequestering carbon to mitigate climate change. Whether established and managed by state agencies, local communities, indigenous peoples, or the private sector, protected areas can reduce deforestation and greenhouse gas (GHG) emissions from land conversion and protect carbon stocks. Terrestrial and oceanic ecosystems serve as major carbon stores and sinks. Forests, for example, cover some 30% of the world's land area but store 50% of the terrestrial carbon in soil and aboveground biomass, while grasslands are estimated to hold 10–30% of the global soil carbon (Parish et al., 2008). Wetlands are especially important as carbon stores. Even though peat lands cover only 3% of the global land area, they contain approximately 30% of all the carbon on land, equivalent to 75% of all atmospheric carbon and twice the carbon stock in the global forest biomass (de Groot et al., 2012). Coastal and marine habitats, especially salt marshes, mangroves, kelp, and sea grass beds, are also important carbon sinks, sequestering carbon more efficiently than terrestrial ecosystems of equivalent area (Laffoley & Grimsditch, 2009).

Unfortunately, land conversion, drainage for agriculture, or forestry can turn natural habitats from carbon stores and sinks to a carbon source. Land conversion is estimated to be responsible for 20% of GHG emissions, while carbon emissions from peat land drainage, fires, and conversion are approximately 3 billion tons per year, which equates to more than 10% of the global fossil fuel emissions (de Groot et al., 2012). Thus conservation of natural ecosystems is a key strategy for climate change mitigation, highlighting the importance of the global protected area network. A conservative estimate suggests that protected areas globally store over 312 billion tons of carbon or 15% of the terrestrial carbon stock, but the degree to which carbon stocks are protected varies among regions and under different management regimes (Campbell et al., 2008; Kapos et al., 2008). National- and state-managed protected areas and indigenous reserves in the Brazilian Amazon, for example, now cover more than 300,000 km^2, protect a carbon stock of 4.5 billion tons, and are likely to prevent an estimated 670,000 km^2 of deforestation by 2050, representing 1.8 billion tons of avoided carbon emissions (World Bank, 2010). Elsewhere, protected areas have been shown to be effective carbon stores (Table 3.1). Assessing where areas of high carbon value overlap with sites of high

Table 3.1 **Examples of benefits afforded by protected areas through ecosystem services in addition to biodiversity conservation**

Benefit	Role of protected area	Protected area habitat type	Examples
Carbon storage	Carbon store	Forests	The Amazon Region Protected Area network (300,000 km^2 of state, provincial, and indigenous reserves) protects an estimated 4.5 billion tons of carbon stock. Reduced emissions estimated at 1.8 billion tons of carbon
			Protected areas in Bolivia, Venezuela, and Mexico contain 250,000 km^2 of forest and are estimated to store over 4 billion tons of carbon, worth US$39–87 billion in terms of global damage costs avoided
	Carbon sink and sequestration	Peat lands	Peat lands globally store 550 Gt of carbon, twice as much as the total global forest biomass. These high carbon values justify the restoration of peat land protected areas in Belarus and swamp forest in Sebangau National Park, Indonesia
		Forest	Uganda. Estimated value of carbon sequestration in national park system is US$17.4 million per year
Water flow and water quality	Watershed protection	Boreal forest, Alpine	The Banff National Park in Canada protects the Bow River Basin, the most highly populated watershed in southern Alberta, providing clean drinking water to 1.2 million people
		Wetlands	Clearance of invasive alien species and restoration of a degraded wetland at Riverside California cost only US$2 million compared to US$20 million to build a conventional denitrification plant to improve water quality
Buffering against tidal waves and storm surges	Physical barrier against waves and coastal erosion	Mangroves, barrier islands, coral reefs, sand dunes	The Sundarbans mangroves in Bangladesh and India stabilize coastlines and buffer inland areas from cyclones
			In Guinea Bissau, the value of the protection against erosion provided by mangroves in the Rio Cacheu park is estimated at €2.7 million per year
	Providing overspill space for tidal surges	Coastal marshes	The Black River Lower Morass, the largest freshwater wetland ecosystem in Jamaica, acts as an overspill area for river floodwaters and incursions by the sea
			Coastal wetlands in the US provide storm protection services valued at US$23.2 billion a year
			Japan is restoring coastal forests to provide coastal protection in areas affected by the Sendai earthquake and tsunami

Protection against flooding	Flood attenuation	Marshes, coastal wetlands, peat bogs, natural lakes	The Whangamarino bog and swamp complex on North Island, New Zealand, plays a significant role in flood control and sediment trapping. Flood control is estimated at NZ$601,037 per year (2003 prices), but in 1998, flood prevention was valued at NZ$4 million
	Absorbing and reducing water flow	Riparian forests	Parana river forests in Argentina—protected areas as part of flood control strategies
Food security	Maintaining fish nurseries and feeding grounds	Coral reefs, mangroves	Well-managed reefs in the Indian and Pacific Oceans can provide between 5 and 15 tons of seafood per km^2 per year Mangrove-related fish and crab species account for 32% of small-scale fisheries in the Gulf of Mexico with mangroves valued at US$37,500 per km^2 per year
		Freshwater lake	Cambodia's inland fisheries are valued at $500 m annually with 60% coming from the Tonlé Sap lake, a UNESCO Man and the Biosphere Reserve
	Maintaining drought-resistant plants	All dry land habitats	In Mali, protected areas are recognized as important reservoirs of drought-resistant species
	Water for irrigated agriculture	Forest	Ankarafantsika National Park in Madagascar protects water supplies to the rice-producing region of Marovoay plains and the Lac Alaotra watershed, a Ramsar site, serves about 800 km^2 of rice farms
Energy	Protecting water flow, quality, and quantity	Paramos	Los Nevados and other high-altitude parks in Colombia safeguard water supplies to downstream hydroplants supplying 60% of Colombia's power supply In Peru, 60% of hydropower comes from rivers in protected areas, a service valued at US$320 million per year

Source: Adapted from Stolton & Dudley (2010) and Lopoukhine et al. (2012).

biodiversity could prove useful in targeting priority areas to establish new protected areas in line with Target 11 (Woodley et al., 2012).

Well-managed protected areas can also play a critical role in adaptation strategies, reducing the vulnerability of local communities to the impacts of climate change, protecting watersheds and soils, shielding communities from natural disasters, sustaining fisheries and other food sources, and maintaining reservoirs of wild crop relatives to enhance agricultural productivity and crop resilience (World Bank, 2010; Dudley et al., 2010; Davies et al., 2012; MacKinnon, 2012). Both the Conventions on Climate Change (UNFCCC) and the CBD now recognize that protected areas and other ecosystem-based approaches can play a key role in local and national strategies to address the causes of climate change and help societies to respond, and adapt, to the changes that are occurring.

Protected areas securing human welfare

Higher temperatures and more erratic rainfall patterns associated with climate change will affect the supply of, and demand for, water resources as well as environmental flows in many regions. Growing concern over water scarcity provides a powerful argument for protection and improved management of natural habitats to maintain water supplies for agriculture, domestic use, and hydropower (World Bank, 2010; Dudley et al., 2010; Stolton & Dudley, 2010; MacKinnon, 2012). Agriculture is the largest user of freshwater globally, taking more than 50% of the freshwater in many countries with irrigated crop production expected to almost double by 2030 in order to meet global demand (World Bank, 2010). Maintaining natural vegetation is often the cheapest and most effective way of securing water supply and quality. Appreciation of the value of intact forests in maintaining downstream water supplies for agricultural irrigation led to the establishment of the 3,000 km^2 Dumoga Bone National Park in North Sulawesi, Indonesia, and to the expansion of Madagascar's protected area network to 60,000 km^2 (World Bank, 2010). In Venezuela, about 20% of the irrigated lands depend on protected areas for their water supplies, while in the Dominican Republic, the Madre de las Aguas Conservation Area protects the source of 17 rivers that provide water for domestic use and irrigation to over half of the country's population (IUCN-WCPA, 2012).

Municipal water accounts for less than 10% of global water use yet clean drinking water is a critical human need. Over half of the world's population lives in towns and cities, yet one third of this urban population lacks access to clean drinking water (World Bank, 2010). Functioning natural ecosystems can help to maintain water quality and in some circumstances also increase water availability through filtration, groundwater renewal, and maintenance of natural flows. A global study found that one third (33 out of 105) of the world's largest cities including Mumbai, New York, Sofia, Dar es Salaam, Melbourne, Tokyo, and

Sydney receive a significant proportion of their drinking water supplies directly from nearby forest protected areas (Dudley & Stolton, 2003). In Colombia, local mayors recognize the value of the Chingaza and Sumapaz National Parks which safeguard metropolitan water supplies for the eight million residents of Bogota and are working with park authorities to protect and restore *paramos* vegetation in the watershed (Stolton & Dudley, 2010). In South Africa, the recognized value of the mountain parks of the Cape Peninsula and Drakensberg in providing water supplies for Cape Town, Johannesburg, and Durban led to the creation of the Working for Water Programme and considerable annual investments to remove water hungry invasive alien tree species (Pierce et al., 2002).

Protected areas can also serve to reduce risk from natural disasters by maintaining natural ecosystems that buffer against extreme climatic events and natural hazards such as hurricanes and tidal surge (coastal mangroves, coral reefs), flash floods (wetlands, floodplains), and landslides (forests and other native vegetations) (Stolton et al., 2008; Dudley et al., 2013). This role will become increasingly important as environmental degradation continues in the broader landscape. While ecosystem protection alone cannot halt the impacts of more extreme climatic events, protected areas can help to buffer vulnerable communities against all but the most severe flood and tidal events, landslides, and storms (Dudley et al., 2013). Around the world, intact mangroves have been shown to provide protection and reduce the damage caused by tsunamis and hurricanes while also supporting vital fish nurseries on which coastal communities depend. In Sri Lanka, the Muthurajawela protected area affords flood protection valued at over US$5 million per year, while in Switzerland, 17% of the forests are managed to stop avalanches, a service valued at US$2–3.5 billion per year (Dudley et al., 2013). Conversely, the loss of coastal wetlands was identified as a major contributory factor in the scale of impacts from Hurricane Katrina in New Orleans in 2005; as a consequence, a major restoration program is now underway in the Louisiana wetlands (Jenkins et al., 2010). In drought-prone regions, conservation of native biodiversity in protected areas and traditional cultural ecosystems, wild food sources, and drought-resistant crop relatives can help communities to cope with the impacts of drought and desertification (Amend et al., 2008).

Investments in protecting and restoring natural habitats may be more cost-effective for reducing disaster risk than investing in hard infrastructure alone (Talberth et al., 2013). Local communities in Vietnam have been planting and protecting mangrove forests in Vietnam since 1994 as a buffer against storms. An initial investment of US$1.1 million has saved an estimated US$7.3 million per year in sea dyke maintenance and significantly reduced the loss of life and property from Typhoon Wukong in 2000 in comparison with other areas (IFRC, 2002). Similarly, in Argentina, protection of natural riverine forests has been incorporated into flood management strategies to complement early warning systems and hard infrastructure along the Parana River, protecting rich biodiversity in the floodplains as well as human settlements (World Bank, 2010).

Until recently, the value of ecosystem services was rarely taken into account in establishing and resourcing protected area networks. The overall estimated value of the ecosystem services provided by both inland and coastal wetlands typically exceeds that of other ecosystems (de Groot et al., 2012; Russi et al., 2013) yet wetlands are some of the most threatened and least protected habitats. The Target 11 imperative to increase protection of inland waters and marine areas is significant both for biodiversity conservation and human livelihoods, with 250 million people globally who are dependent on small-scale fisheries (FAO, 2004). Coral reefs, for example, cover only 0.2% of the world's oceans, but they harbor about 25% of marine species and sustain the livelihoods of millions of people. Marine protected areas have been shown to rejuvenate depleted fish stocks within a few years when they are managed collaboratively with the resource users, providing valuable fish stocks to replenish adjacent fished areas (Halpern, 2003). Fish diversity and overall production in Mexico's Cabo Pulmo National Park has increased by more than 460% since the establishment of a no-take zone within the park in 1995 (NAWPA, 2012).

Protect, connect, and restore

It is clear that many protected areas can deliver multiple goods and services as well as species conservation. But these benefits depend on areas being well managed and well connected. Natural ecosystems are becoming increasingly fragmented, and many protected areas have become isolated islands within more intensively used production lands. Climate change will require new strategies for conservation and sustainable protected area networks, including improved protected area design, establishment of large conservation areas covering altitudinal gradients, maintaining habitat connectivity in the wider landscape, and encouraging more biodiversity-friendly practices in surrounding landscapes. Maintaining connectivity will be a key strategy to allow plant and animal species to adapt to climate change, especially for wide-ranging and migratory species such as large herbivores and predators at the head of the food chain. Protecting large, connected tracts of natural habitats is also an effective strategy for protecting ecosystem services. Linked protected areas in the Australian Alps, for instance, cover 16,400 km^2 of catchments, which deliver some 30% of the total annual inflows to the Murray–Darling Basin, Australia's food bowl. This water benefits more than 2 million people, with an estimated worth of AU$10 billion per year (IUCN-WCPA, 2012). Water and carbon benefits will also be realized by other large-scale connectivity conservation initiatives such as Yellowstone to Yukon in North America, transnational conservation areas in the Greater Virunga landscape in Central Africa, and the Mesoamerican Biological Corridor (Worboys et al., 2010).

According to the Millennium Ecosystem Assessment (MEA), more than 60% of all ecosystems are already degraded (MEA, 2005). Restoration will become an increasingly important part of biodiversity management, both within protected areas and adjacent lands to maintain connectivity between fragmented habitats. Restoration practices can range from improved protection to allow natural regeneration of habitats, control and management of invasive alien species, and reforestation to active restoration of habitats such as wetlands and reintroduction of species (Keenleyside et al., 2012). Whenever possible, restoration efforts should enhance ecosystem services and benefit local communities without compromising conservation objectives. The US Fish and Wildlife Service, for example, has used carbon offsets to plant more than 8 million native trees within 120 km^2 of wildlife refuges over 10 years. As the forests mature, they are expected to sequester more than 9 million tons of carbon, helping to reduce GHG emissions and restore critical wildlife habitat. Similarly, Parks Canada and the Haida Nation are working together in the Gwaii Haanas National Park to restore streams and old-growth forests to reestablish spawning grounds and salmon populations (NAWPA, 2012). In Indonesia, ecological restoration of previous logging concessions has the potential to significantly enhance the conservation estate (Box 3.1).

Funding protected area expansion and management

Establishing new and better managed protected areas to enhance biodiversity conservation will have significant costs. It is hard to get an accurate figure for current levels of funding for protected areas globally. By far the greatest contribution comes from national government budgets but in developing countries these funds are often supplemented by international donor assistance, bilateral aid, NGO support and loans and grants from multilateral development banks. Investments in protected areas globally (from both national and international funds) have been conservatively estimated at US$6 billion with US$750 million spent annually in tropical countries (Balmford & Whitten, 2003). More recent estimates for a truly representative and effectively managed global protected area system are much higher, ranging from US$34 to US$79 billion annually (Butchart et al., 2012; McCarthy et al., 2012) with US$5.1 billion per year needed to manage the network more effectively (CBD, 2012). However where detailed national assessments have been carried out for more effective management of existing protected areas, funding needs have proven to be surprisingly modest. Studies on sustainable financing for protected areas in 20 countries in Latin America showed that effective management of their protected area systems could be achieved for a mere 0.018% of their collective gross national product (Bovarnick et al., 2010). Nevertheless even an estimated need of US$79 billion per year is considerably less than the estimated annual value of US$4400–5200 billion for multiple goods and services from the global protected area network (Balmford et al., 2002). While the costs of establishing protected areas are high, the benefits they provide to society are even higher.

Box 3.1 Harapan Rainforest: Ecological restoration in Sumatra

Over the last century, lowland forest on Sumatra has dwindled from 160,000 km^2 to a mere 5,000 km^2, and the island's lowland forests are still under threat, especially for conversion to oil palm. Although Sumatra has some important conservation areas, most are in hilly regions with very little lowland forest protected and even national parks and reserves threatened by encroachment and illegal logging. Now, new legislation provides exciting new opportunities for enabling production forest, formerly exploited for logging, to be restored and managed for conservation. Harapan Rainforest, 900 km^2 of commercially valuable lowland forest, is the first former concession to be earmarked for ecosystem restoration. The area is managed by a private partnership consisting of three nongovernmental organizations (NGOs), Burung Indonesia, Birdlife International, and the Royal Society for Protection of Birds with a management license for 100 years.

Harapan Rainforest is at the forefront of one of the most exciting conservation opportunities in Indonesia. Although a former logging concession, the area retains some good quality lowland rainforest, a habitat type that is poorly represented in the formal protected area network. The area supports a rich variety of wildlife, including populations of tigers, clouded leopard, elephants, Malay tapir, 6 species of primates, and at least 235 species of birds including 6 species of hornbills. Expansion of the Harapan model to create conservation areas in other sites of production forest could make a significant contribution to Indonesia's efforts to safeguard livelihoods, forests, and biodiversity and to mitigate climate change through reducing GHG emissions from deforestation and degradation. Harapan Rainforest is not formally designated as a protected area, but its management objectives make it equivalent to IUCN Category VI. It fits well with the concept of other conserved areas protecting important biodiversity as described in Target 11 of the Aichi Targets of the CBD (MacKinnon et al., 2012).

Even so, realizing such large funds will be a challenge. The Global Environment Facility (GEF) is one of the largest funding mechanism for protected areas worldwide, having invested more than US$1.6 billion of GEF money plus an additional US$4.2 billion in cofinancing to 1600 protected areas, covering an area equivalent to the combined area of Greenland and Mongolia (Global Environment Facility, 2013). Additionally, the GEF has been a pioneer in supporting conservation trust funds worldwide, investing more than US$300 million in total in funds that support protected areas in 26 countries, including Bhutan, Côte d'Ivoire, Madagascar, Peru, South Africa, and Tanzania. Under the 4-year GEF 5 program (2010–2014), US$700 million was allocated to improve sustainability of protected area systems, and protected areas are also an important focus for GEF 6. Compared to estimated needs, however, these funds are modest.

It is clear that protected areas can never be adequately funded from conservation funds alone and that new financing will need to be identified and linked to economic benefits and national development agendas. The socioeconomic benefits from the provision of ecosystem services, especially water and carbon, could more than justify the costs of effective conservation and sustainable management of natural ecosystems. The value of ecosystem services in terms of water regulation and supply has been estimated as US$2–3 trillion globally (Costanza et al., 1998). Yet to date, very little of this potential value has been spent on protected areas, but encouragingly, things are changing. Latin America is leading the way in exploring the potential of payment for ecosystem services (PES). PES schemes to compensate protected areas, communities, indigenous peoples, and private landowners for maintaining forests and other water-regulating habitats are now being implemented in Costa Rica, Colombia, Ecuador, Mexico, and Nicaragua. Since 1997, Costa Rica has invested over US$100 million in PES schemes. More than 80% of these payments are supporting conservation in national parks, biological corridors, and strategic water catchments, with each hectare of forest estimated to be worth between US$40 and US$100 for the service provided in protecting watersheds (World Bank, 2010). A conservation fund in Colombia, established with support from The Nature Conservancy, will help to protect rivers and watersheds critical to provision of clean drinking water in Bogotá, the country's capital. It has been estimated that water treatment facilities in Bogotá could save US$4 million per year by proactively investing in watershed protection and keeping sedimentation and agricultural run-off out of the country's rivers. With voluntary contributions from water companies and private industry, including the large brewery Bavaria, the fund is projected to raise US$60 million over the next 10 years to subsidize conservation projects, strengthening protected areas and more ecologically sustainable cattle ranching (Goldman et al., 2010).

Government commitments under the UNFCCC on Reducing Emissions from Deforestation and Forest Degradation (REDD and REDD+) have generated much excitement about the potential availability of REDD funding to secure adequate management for forest protected areas. A whole range of carbon funds and market-based financing mechanisms have been developed for forests, including the BioCarbon Fund, Forest Carbon Partnership Facility, Forest Investment Program, and Pilot Program for Climate Resilience (World Bank). The new carbon trading category for wetlands under the Verified Carbon Standard also provides resources for wetland restoration and conservation. All of these sources could provide opportunities to better protect carbon in protected areas as part of strategies to address climate change. So, why isn't this happening on a much larger scale?

First there is the requirement that carbon funds should buy additional benefits and the misplaced assumption that protected areas are already adequately protected because of their legal status. Unfortunately, many protected areas, especially in the tropics, are suffering forest loss through illegal logging and agricultural encroachment, sometimes ironically to plant oil palm for cleaner biofuels. High-priority sites for tackling deforestation to reduce emissions may not always reflect conservation values, and investments in carbon-rich forests may lead to displacement of pressures to other ecosystems either less carbon-rich forests and protected areas or nonforest ecosystems such as savannas or wetlands (Miles & Kapos, 2008). Moreover, while carbon markets may have potential to promote conservation in less productive lands, prospects are less promising where competing land uses, such as conversion to oil palm plantations, may be highly profitable (Damania et al., 2008). There is often also the additional challenge of ensuring the equitable distribution of REDD revenues from national governments to local communities on whom effective forest protection will depend (Peskett et al., 2008; Sandbrook et al., 2010). Nevertheless, important lessons are being learned from pilot schemes where international NGOs have worked with local communities to protect high-biodiversity forests with funding from voluntary carbon markets. Protected areas such as the Noel Kempff Mercado National Park in Bolivia and the Makira protected area in Madagascar are already benefiting from such initiatives. However, carbon funds and other payments for ecosystem services are unlikely to be the silver bullets for meeting conservation costs or community incentives to reduce pressures on protected areas. Most protected areas will need to rely on a basket of funding sources, including government budgets. Engendering the necessary support for increased funding will depend on greater understanding of the role that protected areas can play as cost-effective, proven, and sustainable solutions to reduce vulnerability to climate change and other environmental challenges at both national and local levels (TEEB, 2009, 2011; MacKinnon et al., 2011; Lopoukhine et al., 2012).

Many protected areas contribute multiple benefits in addition to biodiversity conservation. In New Zealand, for instance, the 220 km^2 Te Papanui Conservation Park in the Lammermoor Range provides the Otago region with essential water flows valued at NZ$93 million per year for urban water supply, NZ$12 million per year for irrigation, and NZ$31 million per year for hydroelectricity (BPL, 2006). The economic and social value of coastal and marine ecosystems within Rio Cacheu park, two thirds of which is mangrove habitat, is estimated at €9 million per year (Binet et al., 2012) for direct uses (artisanal fisheries) and indirect services (coastal protection, water treatment, and carbon sequestration). Economic studies in Madagascar showed that the benefits of conservation and sustainable management of 22,000 km^2 of forests and protected areas far outweigh the costs forgone in agricultural production, with 50% of benefits from watershed protection to maintain water flow and reduced sedimentation in downstream smallholder rice fields (World Bank, 2010). The study led to the decision to expand the protected area estate to 60,000 km^2, including protection of 40,000 km^2 of forest. One of the new protected areas, the Ambositra-Vondrozo

Forest Corridor (COFAV), is recognized for its value in providing water for irrigation, agriculture and hydropower generation, benefits for carbon storage, tourism, health (medicinal plants) and food security as well as protecting some of the island's unique biodiversity. When fully operational, the new national system of protected areas in Madagascar could reduce carbon emissions by some 9 million tons per year.

The costs of expanding protected area networks are high, but many protected areas could be justified for their socioeconomic benefits alone (Dudley et al., 2011). However, not all protected areas can be justified through benefits from water provision or other ecosystem services. It may be a particular challenge for small reserves that protect habitat fragments with populations of threatened species and Alliance for Zero Extinction (AZE) sites. To be able to make the case to policy makers, developers, private sector, and the general public, it will be important to get better valuations of the economic and social benefits generated by individual protected areas and national networks (Kettunen & ten Brink, 2013).

Mainstreaming protected areas into development strategies

Adaptation is becoming an increasingly important part of the development agenda, especially in developing countries most at risk from climate change (World Bank, 2010). Much of the investment for adaptation will be directed to hard infrastructure and technological solutions, but investments in hydropower, irrigation dams, reservoirs, and sea walls can all have high environmental costs due to loss of habitats and disruption of ecological flows. Maintaining natural ecosystems and services through establishing and managing protected areas is a smart investment option, particularly for those countries that have large intact ecosystems, since investments in habitat restoration and/or hard infrastructure are likely to be far more costly (Talberth et al., 2013). Several countries are already taking climate change into account in their development strategies. In Namibia, a review of the agriculture sector emphasized that biodiversity-rich ecosystems are more resilient to climate change and that investments in wildlife conservation and protected areas will bring higher and more reliable economic returns, through tourism, than investments in livestock. Countries as diverse as India and Mexico are already writing protected areas into their climate change strategies and undertaking research to quantify the "carbon capture" potential and provision of ecosystem goods and services by protected area landscapes (MacKinnon et al., 2012). In 2009, Cambio Climático en Áreas Naturales Protegidas in Mexico launched the Climate Change Strategy for Protected Areas based on vulnerability assessments, human communities, and production systems. As a result, Mexico is creating corridors to link protected areas and adapting management to include a mixture of core areas and buffer zones while undertaking important policy discussions to align legal instruments and subsidies from other sectors to better support protected areas (MacKinnon et al., 2012).

Through the CBD, countries have already committed to expand their protected area systems for greater representativeness and effective biodiversity conservation. Making protected areas a key part of national and local responses to climate change and other environmental challenges can help to justify such expansion and reduce rates of deforestation, protect carbon-rich habitats, ensure more sustainable land management, and increase the resilience of vulnerable human communities. Ensuring a more sustainable future will require greater recognition of the role that protected areas can play in underpinning development. The conservation community needs to become much more adept at broadcasting the socioeconomic and biodiversity benefits of protected areas so that we can get swift and effective action on the following issues:

- **More and larger protected areas and buffer zones:** To improve ecosystem resilience particularly in areas with both high biodiversity and high carbon or where ecosystem services are under threat, such as in watersheds, tropical lowland forests, peat lands, mangroves, freshwater and coastal marshes, and sea grass beds. Wetlands and marine ecosystems are especially poorly represented in protected area systems yet both provide multiple benefits for human welfare.
- **Connecting protected areas within landscapes and seascapes:** To expand habitat under some form of conservation management beyond park boundaries into buffer zones, biological corridors, and ecological stepping stones, to protect ecosystem services and build resilience to climate change.
- **Strengthening management of protected areas under the full range of governance types:** To maintain water production and carbon and biodiversity values and reduce degradation and loss through threats such as agricultural encroachment, overexploitation, and invasive alien species.
- **Restoration strategies:** To restore degraded habitats within, and around, protected areas to enhance biodiversity, carbon, and other ecosystem services and restore connectivity.
- **Better integration:** To include protected areas within broader development policies, planning, and programs including climate change and disaster risk reduction strategies and spatial planning, to identify places where protected areas are providing essential ecosystem services and where there are social and economic benefits from incorporating green infrastructure within development plans.
- **Improved valuation of services and benefits from individual sites and protected areas networks:** To underpin arguments for strengthened support and innovative financing strategies for protected area networks, including payments for ecosystem services, additional government budgets and financing through major development projects and biodiversity offsets.

References

Amend, T. et al., 2008. *Protected landscapes and agrobiodiversity values*, Gland: IUCN & GTZ.

Balmford, A. & Whitten, T., 2003. Who should pay for tropical conservation and how could the costs be met. *Oryx*, 37, p. 238–250.

Balmford, A. et al., 2002. Economic reasons for conserving wild nature. *Science*, 297(5583), p. 950–953.

Bellard, C. et al., 2012. Impacts of climate change on the future of biodiversity. *Ecology Letters*, 15(4), p. 365–377.

Bergengren, J. C., Waliser, D. E., & Yung, Y. L., 2011. Ecological sensitivity: A biospheric view of climate change. *Climatic Change*, 107, p. 3–4.

Binet, T., Failler, P., Borot de Battisti, A., & Marechal, J. P., 2012. *Évaluation de la valeur économique et sociale des écosystèmes associés aux AMP de l'Afrique de l'Ouest, Guide Méthodologique*, Dakar: PCRM.

Bovarnick, A., Fernandez Baca, J., Galindo, J., & Negret, H., 2010. *Financial sustainability of protected areas in Latin America and the Caribbean*, New York & Arlington: UNDP & TNC.

BPL, 2006. *Evaluation study of the economic benefits of water in Te Papanui Conservation Park*, Christchurch: BPL.

Bruner, A. G., Gullison, R. E., Rice, R. E., & da Fonseca, G. A. B., 2001. Effectiveness of parks in protecting tropical biodiversity. *Science*, 291(125), p. 125–128.

Butchart, S. H. M. et al., 2012. Protecting important sites for biodiversity contributes to meeting global conservation targets. *PLoS ONE*, 7(3), p. e32529.

Campbell, A. et al., 2008. *Carbon emissions from forest loss in protected areas*, Cambridge: UNEP-WCMC.

CBD, 2012. *Report of the high level panel on global assessment of resources for implementing the Strategic Plan for Biodiversity 2011–2020 (UNEP/CBD/COP/11/INF/20)*, Montreal: Secretariat of the CBD.

Costanza, R. et al., 1998. The value of the world's ecosystem services and natural capital. *Ecological Economics*, 25(1), p. 3–15.

Damania, R. et al., 2008. *A future for wild tigers*, Washington, DC: World Bank.

Davies, J. et al., 2012. *Conserving dryland biodiversity*, Nairobi: IUCN, UNEP-WCMC & UNCCD.

Dudley, N. & Stolton, S., 2003. *Arguments for protected areas: Multiple benefits for conservation and use*, London: Earthscan.

Dudley, N. et al., 2010. *Natural solutions: Protected areas helping people cope with climate change*, Gland, Washington, DC & New York: IUCN-WCPA, TNC, UNDP, WCS, World Bank & WWF.

Dudley, N. et al., 2011. National parks with benefits: How protecting the planet's biodiversity also provides ecosystem services. *Solutions*, 2(6), p. 87–95.

Dudley, N., Krueger, L., MacKinnon, K., & Stolton, S., 2012. Ensuring that protected areas play an effective role in mitigating climate change. In: E. A. Beever & J. L. Belant, eds. *Ecological consequences of climate change: Mechanisms, conservation and management*. Boca Raton: Taylor & Francis Group, p. 237–261.

Dudley, N., MacKinnon, K., & Stolton, S., 2013. Reducing vulnerability: The role of protected areas in mitigating natural disasters. In: F. G. Renaud, K. Sudmeier-Rieux, & M. Estrella,

eds. *The role of ecosystems for disaster risk reduction*. Tokyo: United Nations University Press, p. 371–389.
FAO, 2004. *The state of world fisheries and aquaculture*, Rome: FAO.
Gaines, S. D. et al., 2010. Evolving science of marine reserves: New developments and emerging research frontiers. *Proceedings of the National Academy of Sciences of the United States of America*, 107(43), p. 18251–18255.
Global Environment Facility, 2013. Global Environment Facility. [Online] Available at: http://www.thegef.org (Accessed August 7, 2013).
Goldman, R. L., Benitez, S., Calvache, A., & Ramos, A., 2010. *Water funds: Protecting watersheds for nature and people*, Arlington: TNC.
de Groot, R. et al., 2012. Global estimates of the value of ecosystems and their services in monetary units. *Ecosystem Services*, 1(1), p. 50–61.
Halpern, B. S., 2003. The impact of marine reserves: Do reserves work and does reserve size matter. *Ecological Applications*, 13, p. 117–137.
IFRC, 2002. *Mangrove planting saves lives and money in Vietnam: World disasters report focus on reducing risk*, Geneva: IFRC & RCS.
IUCN-WCPA, 2012. *Natural solutions: Protecting areas maintaining essential water supplies*, Gland & Cambridge: IUCN-WCPA.
Jenkins, W. A., Murray, B. C., Kramer, R. A., & Faulkner, S. P., 2010. Valuing ecosystem services from wetlands restoration in the Mississippi alluvial valley. *Ecological Economics*, 69(5), p. 1051–1061.
Juffe-Bignoli, D. 2014. *Protected Planet Report 2014: Tracking progress towards global targets for protected areas*, Gland & Cambridge: IUCN & UNEP-WCMC.
Kapos, V. et al., 2008. *Carbon and biodiversity: A demonstration atlas*, Cambridge: UNEP-WCMC.
Keenleyside, K. A. et al., 2012. *Ecological restoration for protected areas: Principles, guidelines and best practices*, Gland: IUCN.
Kettunen, M. & ten Brink, P., 2013. *Social and economic benefits of protected areas*, Abingdon: Routledge.
Laffoley, D. & Grimsditch, G., 2009. *The management of natural coastal carbon sinks*, Gland: IUCN.
Lopoukhine, N. et al., 2012. Protected areas: Providing natural solutions to 21st century challenges. *Surveys and Perspectives Integrating Environment and Society* (SAPIENS), 5(2), p. 116–131.
MacKinnon, K., 2012. Biodiversity, poverty and climate change: New challengers and opportunities. In: D. Roe, J. Elliot, C. Sandbrook, & M. Walpole, eds. *Biodiversity conservation and poverty alleviation: Exploring the evidence for a link*. Oxford: Wiley-Blackwell, p. 289–303.
MacKinnon, K., Dudley, N., & Sandwith, T., 2011. Protected areas: Helping people to cope with climate change. *Oryx*, 45(4), p. 461–462.
MacKinnon, K., Dudley, N., & Sandwith, T., 2012. *Putting natural solutions to work: Mainstreaming protected areas in climate change responses*, Bonn: BfN.
McCarthy, D. P. et al., 2012. Financial costs of meeting global biodiversity conservation targets: Current spending and unmet needs. *Science*, 338(6109), p. 946–949.
MEA, 2005. *Ecosystems and human well being: Biodiversity synthesis*, Washington, DC: World Resources Institute.

Miles, L., & Kapos, V., 2008. Reducing greenhouse gas emissions from deforestation and forest degradation: Global land use implications. *Science*, 320, p. 1454–1455.

NAWPA, 2012. *North American protected areas as natural solutions to climate change*, Banff: The North American Intergovernmental Committee on Cooperation for Wilderness and Protected Area Conservation.

Parish, F. et al., 2008. *Assessment on peatlands, biodiversity and climate change: Main report*, Kuala Lumpur & Wageningen: Global Environment Centre & Wetlands International.

Peskett, L., Huberman, D., Bowen-Jones, E., & Brown, J., 2008. *Making REDD work for the poor*, London: ODI.

Pierce, S. M., Cowling, R. M., Sandwith, T., & MacKinnon, K., 2002. *Mainstreaming biodiversity in development: Case studies from South Africa*, Washington, DC: World Bank.

Russi, D. et al., 2013. *The economics of ecosystems and biodiversity (TEEB) for water and wetlands*, London & Gland: IEEP & Ramsar Secretariat.

Sandbrook, C., Nelson, F., Adams, W. M., & Agrawal, A., 2010. Carbon, forests and the REDD paradox. *Oryx*, 44(3), p. 330–334.

Selig, E. R. & Bruno, J. F., 2010. A global analysis of the effectiveness of marine protected areas in preventing coral loss. *PLoS ONE*, 5(2), e9278.

Stolton, S. & Dudley, N., 2003. *Running pure: The importance of forest protected areas to drinking water*, Gland & Washington, DC: WWF & World Bank.

Stolton, S., Dudley, N., & Randall, J., 2008. *Natural security: Protected areas and hazard mitigation*, Gland: WWF.

Talberth, J. G., Yonavjak, E., & Gartner, L. T., 2013. Green versus gray: Nature's solutions to infrastructure demands. *Solutions*, 4(1).

TEEB, 2009. *TEEB—The economics of ecosystems and biodiversity for national and international policy makers. Summary: Responding to the value of nature*, Wesseling: Welzel+Hardt.

TEEB, 2011. *The economics of ecosystems and biodiversity in national and international policy making*, London/Washington, DC: Earthscan.

UNEP, 2006. *Marine and coastal ecosystems and human well-being: A synthesis report based on the findings of the Millennium Ecosystems Assessment*, Nairobi: UNEP.

Woodley, S. et al., 2012. Meeting Aichi target 11: What does success look like for protected area systems. *PARKS*, 18(1), p. 23–36.

Worboys, G. L., Francis, W. L., & Lockwood, M., 2010. *Connectivity conservation management: A global guide*, London: Earthscan.

World Bank, 2010. *Convenient solutions to an inconvenient truth: Ecosystem based approaches to climate change*, Washington, DC: World Bank.

4

Optimal Protection of the World's Threatened Birds, Mammals, and Amphibians

J. E. M. Watson[1,2,3], *D. B. Segan*[1] *and R. A. Fuller*[2]

[1]Wildlife Conservation Society, Bronx, NY, USA
[2]School of Biological Sciences, University of Queensland, St. Lucia, Queensland, Australia
[3]School of Geography, Planning and Environmental Management, University of Queensland, St. Lucia, Queensland, Australia

Introduction

The establishment of protected areas to conserve ecosystems and species has arguably been the greatest achievement of global conservation. While there has been generally little to celebrate when it has come to achieving targets set out in the Convention on Biological Diversity (CBD) (Butchart et al., 2010), the growth of the global terrestrial protected area estate over the past two decades is certainly one of the success stories (Butchart et al., 2012; Juffe-Bignoli et al., 2014). In the last two decades, the global protected area estate has doubled in size and now covers approximately 12.5% of the Earth's terrestrial area (Watson et al., 2014). In 2010, the 193 parties of the CBD adopted a new strategic plan, including 20 targets to be met by 2020 (CBD, 2011). Two of these targets relate specifically to conserving species and sites: (i) preventing the extinction of known threatened species and improving and sustaining their conservation status (Target 12) and (ii) effectively managing and expanding protected areas to cover 17% of terrestrial and inland water areas (and 10% of coastal and marine areas), "especially areas of particular importance for biodiversity" (Target 11) (CBD, 2011; McCarthy et al., 2012).

Protected Areas: Are They Safeguarding Biodiversity?, First Edition.
Edited by Lucas N. Joppa, Jonathan E. M. Baillie and John G. Robinson.

A major criticism of the protected area targets outlined in past CBD strategic plans is that they were set without an explicit ecological foundation (Soule & Sanjayan, 1998; Tear et al., 2005; Watson et al., 2008). The 2010 strategic plan is no exception, and while the current Target 11 requires greater protected area coverage than the previous target (compare 17% with 10%) and emphasizes representativeness, it is not explicitly guided by scientific principles of what is needed to conserve biodiversity. Moreover, the new targets are not set on a platform of success in meeting past targets. When the CBD treaty negotiations in Nagoya were taking place in the final months of 2010, the only published science on protected area coverage for threatened species that could be used by policy makers at a global scale was completed in 2004 (based on species and protected area data compiled prior to 2000; Rodrigues et al., 2004a, b). This is despite the fact that since 2000 (when the 2010 strategic plan was accepted) there has been a significant increase in the protected area estate globally (Jenkins & Joppa, 2009, IUCN and UNEP-WCMC, 2012) and also a dramatic increase in the accuracy of distribution maps for threatened species. At the 2010 CBD Conference of the Parties (COP) negotiations, policy makers had no analysis in front of them on 2010 levels of protection for threatened species, or any analysis on the level of protection required for the adequate protection of threatened species. Knowledge on both how well countries are meeting CBD strategic plan targets and what is needed to adequately protect species is clearly important for those who are developing new CBD targets as there is a fine balance between setting high, yet achievable targets versus unrealistic, unachievable targets or targets that can too easily be achieved and have little overall conservation benefit.

In this chapter, we attempt to fill a part of this knowledge vacuum. First, we provide an update on how well threatened bird, mammal, and amphibian (hereafter BAM) species are represented in the global protected area estate. We chose these three taxonomic groups because there has been a complete assessment of the extinction risk of all species for the IUCN Red List. We compare protected area coverage for BAM species with the last available gap analysis conducted by Rodrigues et al. in 2004. We then identify expansion priorities for the global protected area estate to achieve two different objectives. The first objective was to identify which additional areas of the planet need to be protected to represent each threatened species at least once, that is, to transition to a global protected area estate that has no "gap species." The concept of "gap species" originates from gap analysis, a spatially explicit framework for evaluating the coverage of species and ecosystems with respect to tenure regime (Scott et al., 1993). "Gaps" refer to species or ecosystems that are underrepresented or entirely missing from a tenure regime. Here, we use "gap species" to refer to species that are entirely absent from the global protected area estate. While protecting all current gap species is a symbolic achievement, we recognize that simply representing a small portion of the range of each threatened terrestrial vertebrate species will not achieve long-term conservation for the vast majority of species. The second objective was to determine how much land needs to be protected to achieve persistence targets for BAM threatened species. We set persistence targets using the methodology developed by Rodrigues et al. (2004b) and subsequently applied by others (e.g., Kark et al., 2009; Watson et al., 2010; Venter et al., 2014) based on a scaled fraction of geographic range size. In our conclusion, we discuss the implications of these findings for the 2010 Aichi targets and the wider CBD strategic plan more generally.

Materials and methods

To measure the coverage afforded by protected areas to the 4118 threatened species under consideration, two forms of spatial data were required: (i) the spatial locations of existing protected areas and (ii) the spatial distributions of all threatened BAM species.

Protected areas

The current protected area database was developed using the World Database on Protected Areas 2012 (n=192,910; IUCN and UNEP-WCMC, 2012). Following the methods of previous global analyses, we included only protected areas with national designation, excluding all areas protected only by international agreements (Rodrigues et al., 2004a, b; Jenkins & Joppa, 2009). All protected areas with a status other than "designated" were also excluded. For protected areas that lacked a polygonal representation (n=22,343), we created a circular buffer of the appropriate area around its central coordinates and excluded from analysis all those protected areas without a specified areal extent (n=7,003). To prevent overestimation of the areal coverage of the global protected area network, buffered points and polygonal representations of protected areas were merged into a single layer.

Threatened species

Following Rodrigues et al. (2004a, b), we used the IUCN Red List database to identify range maps for BAM species listed as threatened, endangered, or vulnerable, hereafter "threatened species." Our analysis focused on the terrestrial protected area network, so we excluded marine mammals and pelagic or coastal seabirds (SHM Butchart, personal communication). Coastal seabirds were excluded from the analysis because the majority of the mapped range falls in the marine environment. The spatial distribution of each threatened species of mammal and amphibian was available from the IUCN Red List (IUCN, 2012), while the distributions of threatened birds were available from Birdlife International (BirdLife International and NatureServe, 2012). We considered these spatial distributions to represent the "extent of occurrence" rather than the "area of occupancy," despite the fact that some broadly unsuitable occupied areas had been removed from most of the maps (see Gaston & Fuller, 2009). "Extent of occurrence" refers to the entire geographic range bounded by all locations where the species is found, while "area of occupancy" refers to just the subset of area inside the extent of occurrence in which the species actually occurs. We used only those parts of species' distributions where they are native, reintroduced or of uncertain origin, and are considered extant or probably extant. We included 1876 amphibians, 1135 birds, and 1107 mammals in our analyses, a total of 4118 threatened species.

Gap analysis

We assessed the occurrence of threatened species within protected areas using a representation target and a persistence target. The representation target was achieved if any portion of the species' distribution overlapped with the protected area network or if a single planning unit was included in the future network that contained any portion of the species' range. The persistence target was defined based on the methodology developed by Rodrigues et al. (2004a, b) and comprised a scaled fraction of the overall geographic range size for a species, with the target increasing for more narrowly distributed species. Complete (i.e., 100%) coverage by protected areas was required for species with a geographic range of less than 1000 km^2. For wide-ranging species (>250,000 km^2), the target was reduced to 10% coverage, and where geographic range size was intermediate between these extremes, the target was log-linearly interpolated (Figure 4.1).

Spatial prioritization analysis

To explore future scenarios for the growth of the global protected area network using the objectives outlined previously, we used the freely available software Marxan (Ball et al., 2009). This software has been widely used for prioritizing conservation areas (Carwardine et al., 2008; Smith et al., 2008), and it uses a simulated annealing algorithm to select multiple near-optimal sets of areas that meet the prespecified species targets while minimizing overall cost (Ball et al., 2009).

Dividing the terrestrial surface of the Earth into 30×30 km grid cells ($n = 164{,}880$), we undertook two separate spatial prioritization analyses to explore the extent and location of land required to (i) capture one representation of all threatened species in protected areas and (ii) meet the threatened species persistence targets outlined previously. To account for financial costs in the analysis, we used a dataset related to the economic opportunity costs of conservation. Agricultural expansion is the greatest single cause of habitat fragmentation on Earth and we used an adapted version of a recent agricultural opportunity cost dataset (see Naidoo & Iwamura, 2007 for a full description of the dataset). These data do not represent the absolute cost of implementing a protected area but

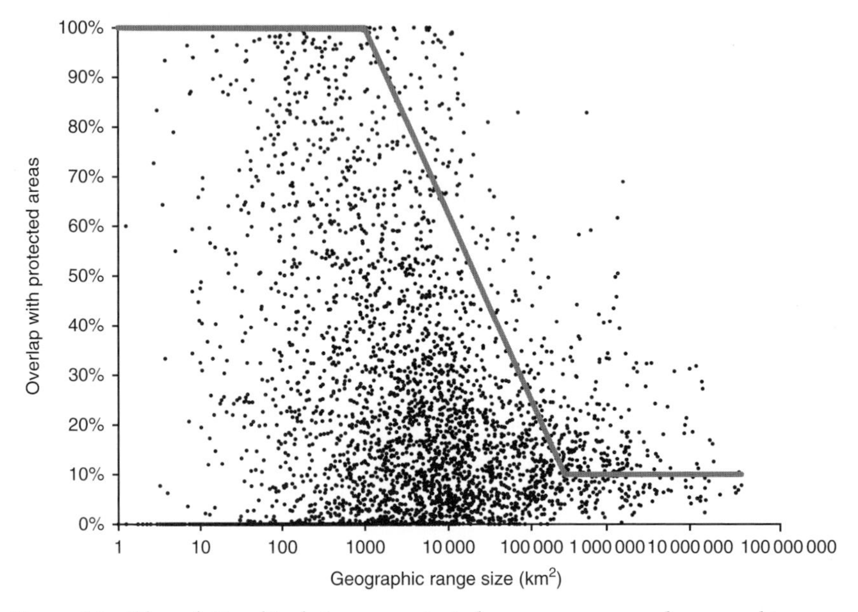

Figure 4.1 **The relationship between protected area coverage and geographic range size for threatened bird, amphibian, and mammal species (*n* = 4118). The gray line represents the persistence target for each species developed by Rodrigues et al. (2004a, b), based on the scaled fraction of range size. Source: Data from Rodrigues et al. (2004a), WDPA (2010), IUCN (2012), and BirdLife International and NatureServe (2012)**

represent the opportunity cost of a protected area that could instead have been used for agriculture. We calculated the area-weighted total cost for each planning unit and treated this as the opportunity cost of establishing a protected area in that planning unit. We applied a fixed transaction cost of $10,000 per 10 km^2 to reflect administrative cost of acquiring areas selected (Fuller et al., 2010). The area-based transaction cost was also incorporated into the estimated opportunity cost of the existing protected area network to facilitate comparison between the current cost of the network and the cost required to achieve the specified objectives. We performed 100 runs for each scenario and chose the solution that achieved the conservation target at the lowest overall cost.

How well represented are threatened birds, mammals, and amphibians in the current PA system?

Of the 4118 threatened species included in this analysis, 717 (17.4%) do not occur in any protected area (Table 4.1). Amphibians had the highest proportion of gap species (22.4%), followed by birds (14.7%) and mammals (11.8%), and this is probably

Table 4.1 **The number of gap species and the number of species that met their representation target (see text) by taxonomic group**

	Number of species	Number of gap species (%)	Number of gap species Rodrigues et al. (2004a, b) (%)*	Number of species that meet persistence targets (%)
Amphibians	1877	420 (22.4)	422 (27.3)	126 (6.7)
Birds	1135	167 (14.7)	244 (20.8)	159 (14.0)
Mammals	1106	130 (11.8)	153 (14.4)	239 (21.6)
Total	4118	717 (17.4)	819 (21.7)	524 (12.7)

Source: Data from WDPA (2010) and IUCN (2012).
*NB total number of species entered into the analysis differs from the present study.

an artifact of the difference in average range between taxonomic groups. Proportional coverage by protected areas declined with increasing threatened species range size (Figure 4.1), although species with the smallest geographic range sizes had both large numbers of species with large levels of protection and large numbers of gap species (Figure 4.1). As range size increases, there was a general decrease in the number of gap species and also a decrease in the proportion of species with large levels of protection. Surprisingly, some species with geographic range sizes exceeding 100,000 km^2 were still completely unprotected by the protected area estate.

While it is impossible to construct a one-to-one comparison with the original Rodrigues et al. (2004a, b) gap analysis (as individual species data and protected area data are not available from that time frame anymore), we were able to assess how things have changed at the group level. Even with the large gains in protected areas over the past decade, we discovered that the numbers of gap species for birds, amphibians, and mammals have declined by only around 4%, with birds having the greatest reductions in gap species and mammals the least (Table 4.1).

Sufficient protected area coverage to meet the persistence target was achieved for only 524 (12.7%) species (Table 4.1). This varied among taxa with 239 mammal species (21.6%), meeting the persistence target, 126 amphibians (6.7%), and 159 birds (14.0%). In contrast, Rodrigues et al. (2004a) concluded that 11% of threatened species met the persistence target, suggesting a slight increase in the levels of protection of threatened species over the past decade.

What is needed to optimally protect birds, amphibians, and mammals, and where are these priority landscapes?

If new sites are optimally placed, some protection for all of the world's threatened birds, amphibians, and mammals could be achieved with only 13.1% of the world's terrestrial surface, an expansion of just 0.2% of the Earth's land surface to the

Table 4.2 **Two different scenarios for the amount of land requiring protection in order to achieve threatened species representation and persistence targets: (1) achieving coverage by at least one protected area for each threatened mammal, bird, and amphibian species and (2) achieving minimum persistence targets for each threatened mammal, bird, and amphibian species (defined by Rodrigues et al., 2004a, b)**

	Current	Minimum representation of threatened species	Minimum persistence of threatened species
Area required (km²)	17,046,481	17,316,582	24,269,145
Area required (% of terrestrial surface)	12.9	13.1	18.4
Number of ecoregions receiving 17% coverage (%)	273 (33.3)	316 (38.5)	517 (63.0)
Number of threatened species achieving one representation (%)	3401 (82.6)	4118 (100)	4118 (100)
Number of threatened species meeting persistence targets (%)	524 (12.7)	755 (18.3)	4118 (100)

Source: Data from WDPA (2010) and IUCN (2012).

current global protected area coverage. Expansion of the protected area system to achieve the more stringent persistence targets would require protection of 18.4% of the Earth's surface with Central Asia, Northern South America, Central America, and Madagascar emerging as key areas in which protection would need to be focused (Table 4.2 and Figure 4.2).

The burden of expanding the global protected area estate to achieve these goals is not shared equitably among countries with Mesoamerica, the Andes, Eastern Africa, and Southeast Asian countries needing to significantly expand their protected area estates (Figure 4.3). It is also clear that the distribution of this burden will also have a disproportionate impact on lower-income countries around the world.

Discussion

By abating the processes that harm biodiversity, protected areas have great potential to slow or stop the decline of threatened species (Taylor et al., 2011; Watson et al., 2014). However, our analysis has revealed a significant number of threatened species globally that remain wholly unprotected (Table 4.1). Fully 717 or 17.4% of threatened BAM species are not represented in the protected are network at all;

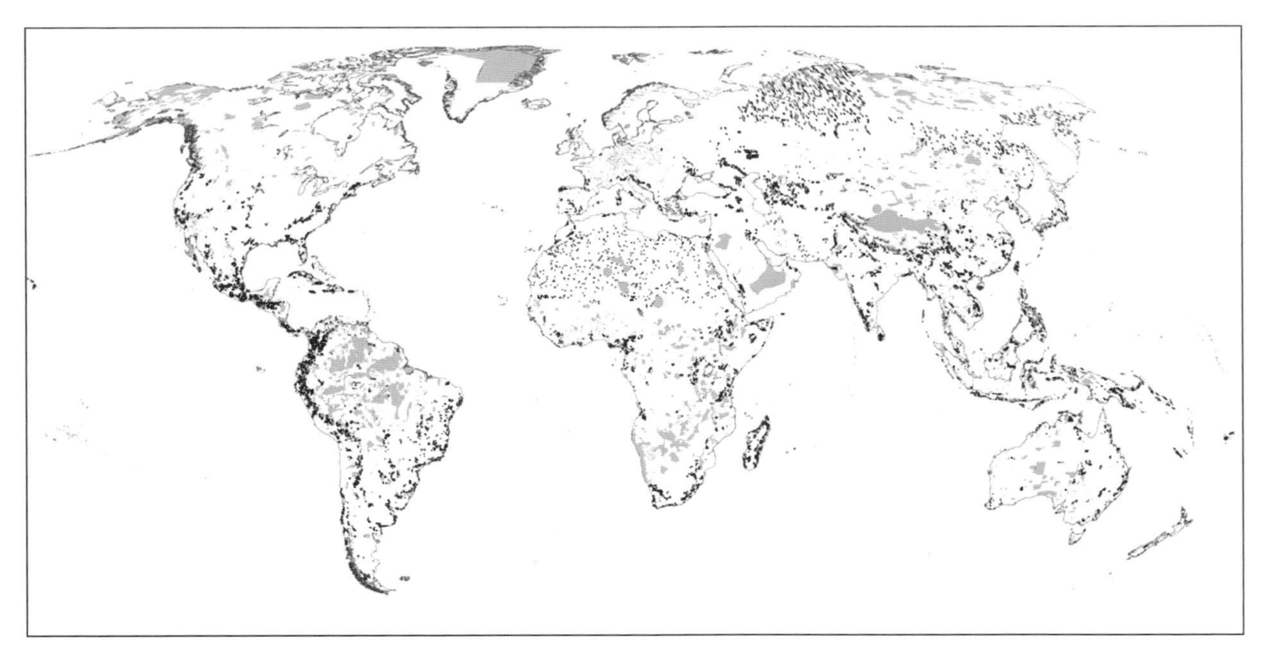

Figure 4.2 **Priority additions (black) to the global protected area estate that capture underrepresented threatened species' distributions. This analysis starts with the existing protected area system and adds new sites that contribute most to meeting persistence targets (see text for details). Gray areas represent existing protected areas. Source: Data from WDPA (2012), IUCN (2012), and BirdLife International and NatureServe (2012)**

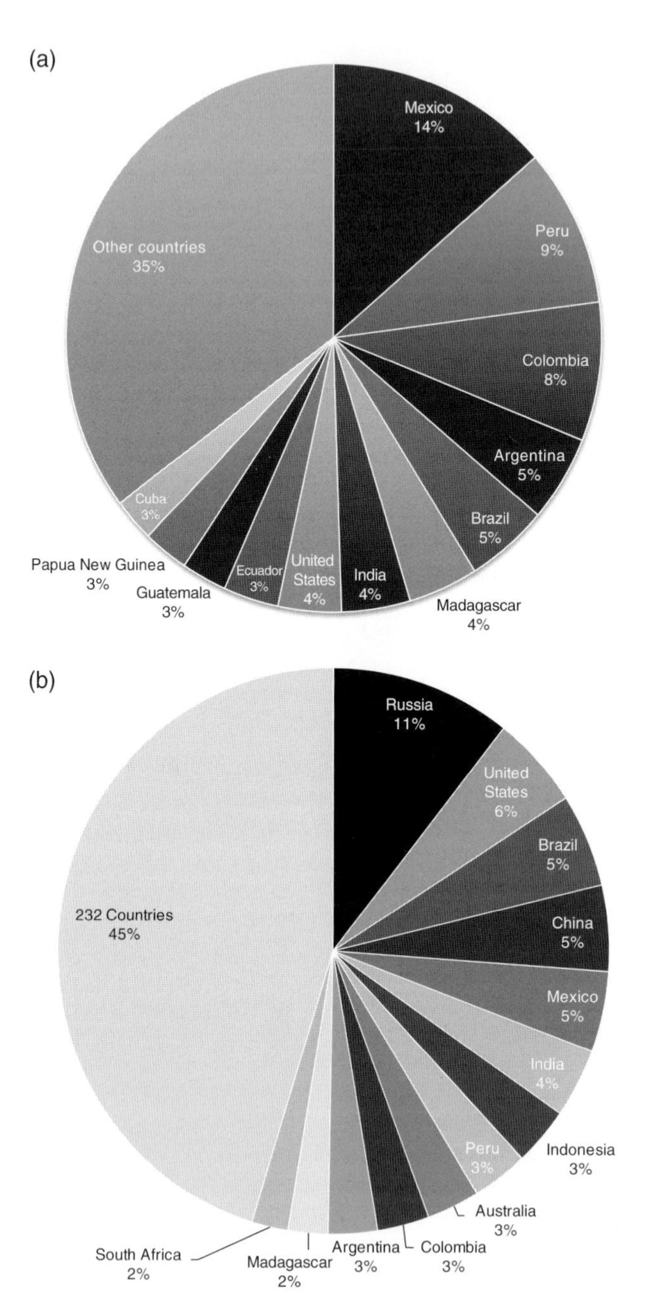

Figure 4.3 **Proportion of new protected areas in an efficiently designed protected area network that lies within the borders of each country and achieves (a) representation of all threatened BAM species, (where total area added is equal to 0.2% of the terrestrial area on Earth) and (b) persistence targets for all threatened BAM species (where total area added is equal to 5.5% of the terrestrial area on Earth). Source: Data from WDPA (2010) and IUCN (2012)**

and 3594 (87.3%) of BAM species are not protected at a level that achieves conservative persistence targets based on range size. Clearly, there is still a need to expand the protected area estate and focus this protection on species objectives and not on static area targets.

We show that an expansion of the terrestrial protected area estate of just 0.2% (from 12.9 to 13.1%) could achieve the goal of representing every threatened terrestrial BAM species at least once. A protected area estate that represents every BAM threatened species is a symbolic achievement, and we believe a strategically targeted expansion of the network could achieve this at relatively little overall cost. If well managed and strategically placed, this expansion will lead to an overall protected area estate that may slow the number of species going extinct in the wild as small populations may be able to hold on inside protected areas, but it would probably do little to halt the decline of those threatened species that have a significant distribution outside protected areas. However, we show that it would require an expansion of 5.5% (from 12.9 to 18.4%) of the Earth's surface for all BAM threatened species to achieve persistence-based targets, assuming all current protected areas are retained and contain intact habitat. The current protected area network, which covers 12.9% of the Earth's terrestrial surface, meets persistence target levels for less than 15% of the 4118 BAM species. With this in mind, it is impressive that with the strategic addition of just 5.5% of the terrestrial surface, we could bring all 4118 threatened BAM species to a persistence level of coverage within the network. This figure is only slightly larger than the current 17% CBD strategic target (CBD, 2011).

The burden of establishing a protected area network that achieves representation or persistence-based targets for all threatened species will not be uniformly shared among countries. Our analysis shows that some countries will likely be asked to play a greater role in establishing and managing new or current protected areas, while little will be asked of other countries. There are two reasons for this inequity. First, threatened species are not uniformly distributed across the globe (Myers et al., 2000), and efficient protection will target locations where multiple species can be protected in a complementary fashion. Second, we aimed to minimize the total opportunity cost of meeting threatened species targets. Attempting to minimize the aggregate opportunity cost across all nations means that we prioritized those areas for protection toward countries with lower opportunity cost. While this minimizes the global cost, it also directs new protected areas disproportionately in countries with low opportunity costs. The result is a protected area network that is economically efficient but is not equitable with respect to either the proportion of area required or the forgone opportunity cost at the country level. While the network we identify may represent an ecologically and economically efficient protected area network, asking individual countries to establish additional protected areas that account for a large proportion of the land area of their country (Figure 4.3) is problematic. Rather than viewing the protected area network configuration that minimizes global opportunity cost as an end point, we view the achievement of an efficient solution as a starting point for further negotiation between countries on the funding of the establishment and management of protected areas. An allocation framework could be developed where a global institution such as the Global Environment Facility (GEF) can facilitate transfers from nations that are being asked to protect less of their land area to countries that are asked to set aside greater portions of their land area, much the same way the Clean Development Mechanism works for the United Nations Framework Convention on Climate Change.

An important caveat in our analysis is that we do not assess the effectiveness of protected areas but simply the amount of geographical overlap between protected areas and threatened species' distributions. The effectiveness of any protected area depends crucially on local circumstances and the amount of investment in management (Brandon et al., 1998), its size (Maiorano et al., 2006) and the level of "connectivity" between protected areas—many protected areas are too poorly managed, too small, or isolated to effectively conserve a population of a threatened species for the long term (Bruner et al., 2001; Watson et al., 2001; Clark et al., 2013). Moreover, as only 40% of BAM threatened species have greater than 20% of their geographic range protected, the importance of management activities outside protected areas cannot be understated. The CBD 2010 strategic plan explicitly recognizes this importance in Target 7 which states: "By 2020 areas under agriculture, aquaculture and forestry are managed sustainably, ensuring conservation of biodiversity" (CBD, 2011). The complementary nature of Target 7 to achieving the objectives of Targets 11 and 12 and dependence of Targets 11 and 12 on the effective implementation of Target 7 necessitate that the broad criteria set forth in the CBD strategic plan be both interpreted and implemented with respect to the specific requirements of threatened species. Ignoring those requirements could result in mixed use landscapes that support only a subset of the most common, adaptable, generalist species and as such fail to represent the full suite of biodiversity or to complement the role that protected areas play in ensuring that threatened species are not driven to extinction.

It is important to note that we considered relative cost based on forgone agricultural opportunity, rather than management costs or opportunity costs for other land uses such as mining or timber (McCarthy et al., 2012). We also recognize the limitations of the biodiversity data used in the analysis. Commission errors, in which a species is mapped as present in a protected area but in reality is absent, are likely to prevail over omission errors in the geographic range dataset, because many factors other than major habitat preferences further limit the distributions of species within their extent of occurrence (Rondinini et al., 2006; Gaston & Fuller, 2009; Venter et al., 2014). This is likely to impact our results in two ways. First, it is likely to cause us to underestimate the number of species insufficiently covered by the protected area network, so our results should be interpreted as conservative in this regard. A second likely effect on our results is an underestimation of the amount of additional land needed to improve representation of gap species by the protected area system. This occurs because coarse areas of occupancy maps overestimate the extent to which species' distributions overlap with each other and thus the ability of any given protected area to simultaneously capture multiple species. We therefore caution that the results shown in Figure 4.2 illustrate the required proportional regional protected area coverage only—they may not be suitable for informing the actual location of area for protected area expansion, as this requires fine-scale socioeconomic and biological data for each country or region (see, e.g., Butchart et al., 2012).

Conclusion

Protected areas are the cornerstone of effective conservation of threatened species (Brooks et al., 2004; Butchart et al., 2015). However, rates of gazetting new terrestrial protected areas have slowed dramatically (Jenkins & Joppa, 2009; Watson et al., 2014), emphasizing the need for efficiency in future acquisitions. We highlight that the current protected area estate does not adequately represent threatened species and there is a need to expand the protected area estate to meet their needs. We also show there is significant between country differences in the investments required to better represent threatened species in protected areas. These results suggest that while it may be feasible to achieve a global target for threatened species protection, there is an urgent need for new policy mechanisms and incentives that allow low-priority nations to fund higher-priority nations to expand their protected areas well beyond their current levels. Failure to recognize the burden that some countries face and to plan efficiently and coordinate investments will jeopardize the chance to achieve global threatened species protection.

References

Ball IR, Possingham HP, Watts ME. 2009. Marxan and relatives: Software for spatial conservation prioritization. In Moilanen A, Wilson KA, Possingham HP, eds. *Spatial conservation prioritisation: Quantitative methods and computational tools*, pp. 185–195. Oxford, UK: Oxford University Press.

BirdLife International and NatureServe, 2012. *Bird species distribution maps of the world.* Cambridge, UK/Arlington, TX: BirdLife International and NatureServe.

Brandon K, Redford KH, Sanderson SE. 1998. *Parks in peril: People, politics, and protected areas.* Washington, DC: Island Press.

Brooks TM, Bakarr MI, Boucher T, Da Fonseca GAB, Hilton-Taylor C, Hoekstra JM, Moritz T, Olivieri S, Parrish J, Pressey RL; Rodrigues ASL, Sechrest W, Stattersfield A, Strahm W, Stuart SN. 2004. Coverage provided by the global protected-area system: Is it enough? *Bioscience* 54: 1081–1091.

Bruner AG, Gullison RE, Rice RE, da Fonseca GAB. 2001. Effectiveness of parks in protecting tropical biodiversity. *Science* 291: 125–128.

Butchart SHM, Walpole M, Collen B, van Strien A, Scharlemann JPW, Almond REA, Baillie JEM, Bomhard B, Brown C, Bruno J, Carpenter KE, Carr GM, Chanson J, Chenery AM, Csirke J, Davidson NC, Dentener F, Foster M, Galli A, Galloway JN, Genovesi P, Gregory RD, Hockings M, Kapos V, Lamarque J-F, Leverington F, Loh J, McGeoch MA, McRae L, Minasyan A, Morcillo MH, Oldfield TEE, Pauly D, Quader S, Revenga C, Sauer JR, Skolnik B, Spear D, Stanwell-Smith D, Stuart SN, Symes A, Tierney M, Tyrrell TD, Vié J-C, Watson R. 2010. Global biodiversity: Indicators of recent declines. *Science* 328: 1164–1168.

Butchart SH, Scharlemann JP, Evans MI, Quader S, Aricò S, Arinaitwe J, Balman M, Bennun LA, Bertzky B, Besançon C, Boucher TM, Brooks TM, Burfield IJ, Burgess ND, Chan S, Clay RP, Crosby MJ, Davidson NC, De Silva N, Devenish C, Dutson GC, Fernández DF, Fishpool LD, Fitzgerald C, Foster M, Heath MF, Hockings M, Hoffmann M, Knox D, Larsen FW, Lamoreux JF, Loucks C, May I, Millett J, Molloy D, Morling P, Parr M, Ricketts TH, Seddon N, Skolnik B, Stuart SN, Upgren A, Woodley S. 2012. Protecting important sites for biodiversity contributes to meeting global conservation targets. *PLoS ONE* 7: e32529.

Butchart SHM, Clarke M, Smith RJ, Sykes RE, Scharlemann JPW, Harfoot M, Buchanan GM, Angulo A, Balmford A, Bertzky B, Brooks TM, Carpenter KE, Comeros-Raynal MT, Cornell J, Ficetola GF, Fishpool LDC, Fuller RA, Geldmann J, Harwell H, Hilton-Taylor C, Hoffmann M, Joolia A, Joppa L, Kingston N, May I, Milam A, Polidoro B, Ralph G, Richman N, Rondinini C, Segan D, Skolnik B, Spalding M, Stuart SN, Symes A, Taylor J, Visconti P, Watson J, Wood L, Burgess ND. 2015. Shortfalls and solutions for meeting national and global conservation area targets. *Conservation Letters*. doi:10.1111/conl.12158.

Carwardine J, Wilson KA, Watts M, Etter A, Klein CJ, Possingham HP. 2008. Avoiding costly conservation mistakes: The importance of defining actions and costs in spatial priority setting. *PLoS ONE* 3: e2586.

CBD. 2011. CBD, COP Decision X/2: Strategic plan for biodiversity 2011–2020, http://www.cbd.int/decision/cop/?id=12268 (accessed July 18, 2015).

Clark NE, Boakes EH, McGowan PJK, Mace GM, Fuller RA. 2013. Protected areas in South Asia have not prevented habitat loss: A study using historical models of land-use change. *PLoS ONE* 8: e65298.

Fuller RA, McDonald-Madden E, Wilson KA, Carwardine J, Grantham HS, Watson JEM, Klein CJ, Green DC, Possingham HP. 2010. Replacing underperforming protected areas achieves better conservation outcomes. *Nature* 466: 365–367.

Gaston KJ, Fuller RA. 2009. The sizes of species' geographic ranges. *Journal of Applied Ecology* 46: 1–9.

IUCN. 2012. IUCN Red List of Threatened Species. Version 2012.2. www.iucnredlist.org (accessed March 19, 2012).

IUCN and UNEP-WCMC, 2012. The World Database on Protected Areas (WDPA): February 2012. Cambridge, UK: UNEP-WCMC.

Jenkins CN, Joppa L. 2009. Expansion of the global terrestrial protected area system. *Biological Conservation* 142: 2166–2174.

Juffe-Bignoli D, Burgess N, Bingham H, Belle EMS, de Lima MG, Deguignet M, Bertzky B, Milam AN, Martinez-Lopez J, Lewis E, Eassom A, Wicander S, Geldmann J, von Soesbergen A, Arnell AP, O'Connor B, Park S, Shi Y, Danks FS, MacSharry B, Kingston N. 2014. *Protected planet report 2014*. Cambridge, UK: UNEP-WCMC.

Kark S, Levin N, Grantham HS, Possingham HP. 2009. Between-country collaboration and consideration of costs increase conservation planning efficiency in the Mediterranean Basin. *Proceedings of the National Academy of Sciences of the United States of America* 106: 15368–15373.

Maiorano L, Falcucci A, Boitani L. 2006. Gap analysis of terrestrial vertebrates in Italy: Priorities for conservation planning in a human dominated landscape. *Biological Conservation* 133: 455–473.

McCarthy DP, Donald PF, Scharlemann JP, Buchanan GM, Balmford A, Green JM, Bennun LA, Burgess ND, Fishpool LD, Garnett ST, Leonard DL, Maloney RF, Morling P, Schaefer HM, Symes A, Wiedenfeld DA, Butchart SH. 2012. Financial costs of meeting global biodiversity conservation targets: Current spending and unmet needs. *Science* 338: 946–949.

Myers N, Mittermeier RA, Mittermeier CG, da Fonseca GAB, Kent J. 2000. Biodiversity hotspots for conservation priorities. *Nature* 403: 853–858.

Naidoo R, Iwamura T. 2007. Global-scale mapping of economic benefits from agricultural lands: Implications for conservation priorities. *Biological Conservation* 140: 40–49.

Rodrigues ASL, Resit Akçakaya H, Andelman SJ, Bakarr MI, Boitani L, Brooks TM, Chanson JS, Fishpool LDC, Da Fonseca GAB, Gaston KJ, Hoffmann M, Marquet PA, Pilgrim JD, Pressey RL, Schipper J, Sechrest W, Stuart SN, Underhill LG, Waller RW, Watts MEJ, Yan X. 2004a. Global gap analysis: Priority regions for expanding the global protected-area network. *Bioscience* 54: 1092–1100.

Rodrigues ASL, Andelman SJ, Bakarr MI, Boitani L, Brooks TM, Cowling RM, Fishpool LDC, Da Fonseca GAB, Gaston KJ, Hoffmann M, Long JS, Marquet PA, Pilgrim JD, Pressey RL, Schipper J, Sechrest W, Stuart SN, Underhill LG, Waller RW, Watts MEJ, Yan X. 2004b. Effectiveness of the global protected area network in representing species diversity. *Nature* 428: 640–643.

Rondinini C, Wilson KA, Grantham H, Boitani L, Possingham HP. 2006. Tradeoffs of different types of species occurrence data for use in systematic conservation planning. *Ecology Letters* 9: 1136–1145.

Scott JM, Davis F, Csuti B, Noss R, Butterfield B, Groves C, Anderson H, Caicco S, D'Erchia F, Jr, Edwards TC, Ulliman J, Wright RG. 1993. Gap analysis: A geographic approach to protection of biological diversity. *Wildlife Monographs* 123: 3–41.

Smith RJ, Easton J, Nhancale BA, Leader-Williams N, Armstrong AJ, Goodman PS, Culverwell J, Dlamini SD, Loffler L, Matthews WS, Monadjem A, Mulqueeny CM, Ngwenya P, Ntumi CP, Soto B. 2008. Designing a transfrontier conservation landscape for the Maputaland centre of endemism using biodiversity, economic and threat data. *Biological Conservation* 141: 2127–2138.

Soule ME, Sanjayan M. 1998. Conservation targets: Do they help? *Science* 279: 2060–2061.

Taylor M, Sattler PS, Evans M, Fuller RA, Watson JEM, Possingham HP. 2011. What works for threatened species recovery? An empirical evaluation for Australia. *Biodiversity and Conservation* 20: 767–777.

Tear TH, Kareiva P, Angermeier PL, Comer P, Czech B, Kautz R, Landon L, Mehlman D, Murphy K, Ruckelshaus M, Michael Scott J, Wilhere G. 2005. How much is enough? The recurrent problem of setting measurable objectives in conservation. *BioScience* 55: 835–849.

Venter O, Fuller RA, Segan DB, Carwardine J, Brooks T, Butchart SHM, Di Marco M, Iwamura T, Joseph L, O'Grady D. 2014 Targeting global protected area expansion for imperiled biodiversity. *PLoS Biology* 12: e1001891.

Watson J, Freudenberger D, Paull D. 2001. An assessment of the focal species approach for conserving birds in variegated landscapes in southeastern Australia. *Conservation Biology* 15: 1364–1373.

Watson J, Fuller RA, Barr L. 2008. Why are we still using a 'one size fits all' philosophy for systematic reserve planning in Australia? *Pacific Conservation Biology* 14: 233–235.

Watson JEM, Evans MC, Carwardine J, Fuller RA, Joseph L, Segan D, Taylor M, Fensham RJ, Possingham HP. 2010. The capacity of Australia's protected area system to represent threatened species. *Conservation Biology* 25: 324–332.

Watson JEM, Dudley N, Segan DB, Hockings M. 2014. The performance and potential of protected areas. *Nature* 515: 67–73.

WDPA, 2010. *The world database on protected areas (WDPA)*. Gland & Cambridge: IUCN & UNEP-WCMC.

5

Maintaining a Global Data Set on Protected Areas

A. Milam[1,2], S. Kenney[1], D. Juffe-Bignoli[1], B. Bertzky[1,3], C. Corrigan[1], B. MacSharry[1], N. D. Burgess[1,4] and N. Kingston[1]

[1]United Nations Environment Programme World Conservation Monitoring Centre, Cambridge, UK
[2]EcoLogic, LLC, Birmingham, AL, USA
[3]Institute for Environment and Sustainability (IES), European Commission, Joint Research Centre (JRC), Ispra, Italy
[4]Center for Macroecology, Evolution, and Climate, Natural History Museum of Denmark, University of Copenhagen, Copenhagen, Denmark

The need to maintain accurate data on the location, size, type and status of the world's protected areas has become critically important in light of continued species decline and habitat degradation. The World Database on Protected Areas (WDPA) fills this need by storing and making available standardised spatial and attribute information on the global protected area network. When combined with other data sets and databases, such as the Global Database on Protected Area Management Effectiveness (GDPAME), and analysed through Geographic Information Science (GIS), we can advance our understanding of the extent of protected areas as a conservation strategy (Coad et al., 2013). This chapter focuses on the successes and limitations of the WDPA, especially with regard to calculating biodiversity protection indicators. A synopsis on the WDPA, its history, purpose, standards, quality and its viability as a biodiversity protection indicator are given. The results of recent global analyses using the WDPA that aimed to track progress towards Aichi Target 11 are also summarised in this chapter. Finally, challenges of maintaining the WDPA are discussed, and consideration is given to the foundations that underpin it as well as solutions for maintaining its viability in the future.

Protected Areas: Are They Safeguarding Biodiversity?, First Edition.
Edited by Lucas N. Joppa, Jonathan E. M. Baillie and John G. Robinson.

The world database on protected areas

The first record of the world's protected areas originated from resolution 713 of the 1959 United Nations Economic and Social Council (ECOSOC) General Assembly, in which the role of protected areas for their economic, social, scientific and environmental wellbeing was formally recognised, and the United Nations List of National Parks and Equivalent Reserves was established (Deguignet et al., 2014). All iterations of the United Nations List were eventually transcribed into the WDPA in 1981. All subsequent UN Lists became an important part of the WDPA process, and the UN List mandate remains a major impetus for updates. The WDPA has additionally acquired an extensive list of policy decisions and recommendations from the Convention on Biological Diversity (CBD), the International Union for Conservation of Nature (IUCN) World Parks Congress and World Conservation Congress, and other international and regional fora, establishing it as the world's only authoritative database on protected areas. International conventions, regional forums and non-governmental organisations (NGOs) alike rely on the WDPA to track progress towards biodiversity targets such as United Nations Millennium Development Goal 7 and Aichi Target 11 of the CBD (CBD, 2010; Leadly, 2014; UNEP, 2014). Governments report to the WDPA under these international obligations, but they also use the WDPA for a variety of practical purposes including public outreach, comparison with other countries and visualisation of their contribution to the global protected area network (Environment Canada, 2015). A growing number of international companies also use the WDPA for decision making, environmental impact assessments and in setting corporate biodiversity policies (Besancon, 2007). However, the research community makes up the largest user base, analysing the WDPA in a wide range of national- to global-scale studies. Many of these uses aim to inform decision makers who in turn help to better the world's conservation efforts.

The WDPA is a joint product of the United Nations Environmental Programme (UNEP) and IUCN, managed by the UNEP World Conservation Monitoring Centre (UNEP-WCMC), in collaboration with governments, NGOs, academia and industry (UNEP-WCMC, 2015). Data are primarily sourced from governments, with lesser amounts from collaborating NGOs, institutions and individuals (UNEP-WCMC, 2015).

The WDPA is accessible online through www.protectedplanet.net.

Setting a global standard

The WDPA Data Standard was established in 2010 as a global standard for protected areas data after 50 years of data compilation had resulted in an inconsistent and incomplete collection of spatial and aspatial information in the WDPA. The development of a data standard signalled a paradigm shift in WDPA governance from a focus on acquisition of data to a focus on quality.

The WDPA Data Standard establishes the type of sites that can be included in the WDPA, the database schema, the content requirements, metadata and validation procedures. All sites included in the WDPA must be compliant with the definitions of a 'protected area'[1] given by IUCN and the CBD as they are reported by IUCN and the CBD Secretariat as having the same meaning (Lopoukhine & de Souza Dias, 2012). Efforts are also underway to provision the WDPA for the inclusion of 'other effective area-based conservation measures' (OECMs) as described in Aichi Target 11 (CBD, 2010); however, a lack of a definition for OECMs prevents the inclusion of these sites for the time being (Kingston et al., 2015; UNEP-WCMC, 2015). Following from the IUCN definition, the IUCN guidelines on protected area management (Dudley, 2008) are an important part of the WDPA Data Standard with regard to protected area management and governance.

The database is stored as an ESRI file geodatabase with two feature classes: one for storing protected areas with boundary data available and another for storing the coordinates of protected areas without boundary data available. Both feature classes contain all the attributes shown in Figure 5.1 for each protected area. Attributes are categorised as 'Minimum' or 'Complete'. 'Minimum' attributes are a priority for key analyses, such as protected area coverage and growth in protected areas over time, and therefore are a minimum requirement for all data submitted to the WDPA. 'Complete' attributes represent information that is either new to the WDPA standard, and therefore largely incomplete or invalidated, or is supplementary to a minimum attribute, such as 'No Take'. 'Complete' attributes are also important for analysis and reporting on issues such as protected area management, governance and ownership (UNEP-WCMC, 2015).

The WDPA Data Standard serves as a framework for collating protected areas data at global, regional and national scales. Several sub-global initiatives have already adopted it as the core standard for their protected area databases, including the following:

- The Marine Protected Areas in the Atlantic Arc (MAIA) database, led by the Agence des Aires Marines Protégées in France
- The Marine Protected Areas in the Northeast Atlantic database of the Oslo and Paris Convention (OSPAR)
- The Marine Protected Areas in the Mediterranean database (MAPAMED) of the Mediterranean Protected Areas Network (MedPAN)
- The European Common Database on Designated Areas (CDDA) of the European Environment Agency (EEA)
- The North American Environmental Atlas of the Commission for Environmental Cooperation (CEC)
- The Protected Areas in the Arctic database of the biodiversity working group of the Arctic Council
- The Digital Observatory for Protected Areas (DOPA) of the Joint Research Centre of the European Commission (Dubois et al., 2015)

WDPA attributes
Minimum [WDPA ID*] [WDPA PID*] [METADATA ID*] [GIS Area*] [GIS Marine Area*] [PA_DEF*] [VERIF*] Name Country Territory (if applicable) Designation Designation type Marine Reported marine area (km^2) Reported area (km^2)
Complete Status Status year Original name Sub-national location Designation (english) Designation type IUCN category International criteria Governance type Ownership type Management authority Management plan (url) No take No take area

Figure 5.1 **WDPA attributes. *Indicates attributes assigned or calculated by UNEP-WCMC (2015). Source: Data from WDPA (2014)**

Source information is stored separately as a table that can be linked to the feature classes, ensuring authorship, disclaimers and data-specific citations are maintained and transparent to WDPA users. When data is provided to the WDPA, a source document and data contributor agreement serves as a record of what was submitted, when and by whom and ensures agreement to include, but not transfer, intellectual property according to the terms of the WDPA. Statements of compliance help validate the data, ensuring the data provider acknowledges compliance with the protected area definitions and in the future will serve to validate data on OECMs (UNEP-WCMC, 2015).

WDPA quality

Current quality assessment

As the curator of the WDPA, UNEP-WCMC does not currently create or alter boundaries. Therefore, the quality of the WDPA varies according to the quality of the data sets provided by the original sources. While the WDPA Data Standard provides some quality control for new data submissions by ensuring consistency of the data set, data quality is primarily measured and reported on, rather than controlled. Quality in the WDPA is measured for each protected area record and aggregated by country, based on four criteria:

- Ratio of points versus polygons
- Currency of the data sets
- Completeness of the attributes
- Consistency between the Reported Area and GIS Area of the polygon

The ratio of points to polygons is considered the most important metric of WDPA quality. Secondly, the age of the data (as reported by the data provider) provides an indication of quality based on the assumption that a data set will be more accurate the more recently it was created or last updated. Data sets obtained, reviewed or updated within the last 5 years are considered better than data sets 5–10 years old, which are considered better still than data sets that are older than 10 years. The validity of using currency as an indicator of quality is also likely to vary by data set and country. For example, if a data provider lacks the capacity or resources to maintain their data set, even when new protected areas are established, the data in the WDPA is unlikely to change until technical capacity is improved and resource gaps are addressed. Alternatively, old data that was accurate at its creation will still be accurate over time even if not updated. UNEP-WCMC addresses these exceptions by asking data providers and experts to validate the data even if they cannot update the data. These validations allow the WDPA to report when data was last verified, having the same effect as an update on currency. Attribute completeness is an important quality metric as it indicates how fit the data is for analysis. Minimum attributes are weighted more heavily than those that are not required for the WDPA. Discrepancies in the area attributes (reported area versus GIS area) are extrapolated as an indicator of quality on their own. A difference of greater than 5% between the area reported by the data provider (reported area) and the area calculated from the polygon (GIS area) indicates error in the data. These individual measures of content in the database, taken together, give an indication of quality in the WDPA and its suitability for use in various analyses. As the quality assessment is aggregated by country, it also highlights those countries that should be assigned lower confidence levels for national analyses and serves to identify priority countries for update and validation (UNEP-WCMC, 2015).

Future quality assessment

New metrics for assessing quality and enhanced validation procedures are always under development. Integrating these will take time and rigorous consultation. WDPA completeness, for example, is currently only partially measured in the WDPA by quantifying missing attributes and missing boundaries. However, quantifying missing sites altogether is complicated by a number of factors. For example, it is well known that some types of protected areas are under-represented in the WDPA, such as privately protected areas (PPAs) and indigenous and community conserved areas (ICCAs), either because they are not mapped, not defined as protected areas or not recognised by national governments in their national reporting to the WDPA or CBD. The exact number and extent of these areas is unknown. In addition to assessing completeness as a measure of missing sites, it is suspected that there are sites in the WDPA which should not be. For example, national governments may include some sites on their national lists that do not meet the IUCN and CBD definitions of a protected area or have any benefit to biodiversity, such as city parks, building structures and green fields for development and therefore should not be included in the WDPA. However, there is no globally consistent way to suss these spurious sites out without country-level verification, which is further complicated when countries do not adhere to or fully understand the criteria for assessing the qualification of designated area as protected areas. Furthermore, the inclusion of OECMs in the CBD's Aichi Target 11 may take biodiversity protection beyond strictly protected areas (CBD, 2010) or it may just mean the inclusion of missing ICCAs and PPAs. Others have attempted to determine WDPA completeness by comparing WDPA country data sets with national data sets or lists (Visconti et al., 2013). However, the uncertainty about what counts and the inconsistency in national versus global reporting described here render the simple comparison of a national list to the WDPA inadequate to determine completeness. While the quality assurance process will continue to evolve, the primary approach to improving data in the WDPA will remain centred on developing the WDPA Data Standard, providing clear guidance on appropriate uses of the WDPA, transparency on quality assessment, consistent updates and validation.

Updates and validation

At present, UNEP-WCMC actively seeks to update every country in the world every 5 years, accepting frequent ad hoc updates while also setting up formal annual to biannual update cycles under data sharing partnerships. Through the updates process, data providers are asked to review and update or validate all sites in their WDPA subset and to add or remove any sites as appropriate. UNEP-WCMC's internal validation procedure includes a set of over 50 consistency and integrity checks performed on every protected area record provided for update. These checks are listed in the appendix of the WDPA Manual (UNEP-WCMC, 2015). UNEP-WCMC, IUCN and the IUCN World Commission on Protected Areas (WCPA) also periodically conduct validation of data through expert reviews. Expert reviews have occurred in the United Kingdom and Asia with plans for reviews in Europe and North America. In the United Kingdom, an expert working group of the WCPA collaborates with NGOs to ensure all private protected areas are assessed against IUCN guidelines, formatted according to the WDPA Data Standard and included in the WDPA with an IUCN Category. In Asia, IUCN and UNEP-WCMC developed expert review projects with governments to review and improve the state of national protected area data sets in the WDPA. To date, this has involved Indonesia, Thailand, South Korea, China and Japan, with support from the Korea National Park Service and the ASEAN Centre for Biodiversity (Juffe-Bignoli et al., 2014a). Expert reviews are ideal in countries where legacy data in the WDPA is an issue or where updates are difficult to obtain through conventional methods.

Poor data quality and implications for analysis

Some protected area data collected prior to the WDPA Data Standard included little to no attribute information. The persistence of legacy data throughout modern versions of the WDPA is a major contributor to gaps in some of the WDPA, as shown in Figure 5.2.

These data gaps can have significant implications for analysis. Common examples of where gaps in key attributes have presented issues for analysts include gaps in status year, IUCN Category and missing boundaries (Spalding et al., 2013; Visconti et al., 2013; Knowles et al., 2015). Where there are gaps in status year (i.e. the year of establishment) for the majority of sites in a country data set, it is impossible to plot growth in number of protected areas over time accurately for that country. This data gap will also skew global coverage graphs. Analysts have attempted to overcome this challenge in various ways. For example, they may choose to include all sites without a status year value as a separate group or alternatively include them in every year group, as was done for the 2011–2014 Millennium Development Goal analysis and Protected Planet Reports (Bertzky et al., 2012; UNEP, 2014; Juffe-Bignoli et al., 2014a).

Gaps in IUCN Category present a similar challenge for analysis. In 2014, 35% of the sites in the WDPA were missing an IUCN Category (Jonas et al., 2014). Were categories are missing they are often excluded from the analysis (Jenkins & Joppa, 2010; Jenkins et al., 2013). Even where categories are assigned, the interpretation of how the categories are applied can differ between nations.

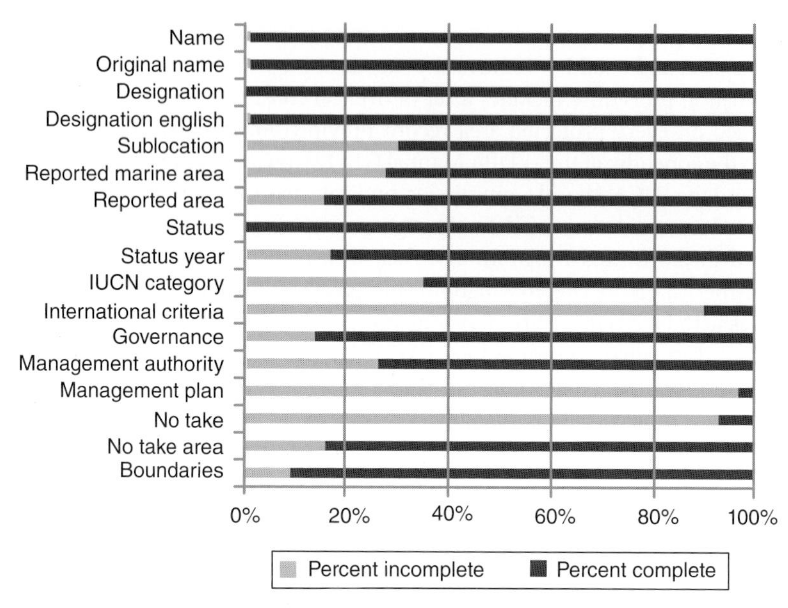

Figure 5.2 **Attribute gaps in the WDPA in August 2014. Source: Data from WDPA (2010)**

International sites such as Ramsar, World Heritage Sites and Man and the Biosphere Reserves do not have IUCN Categories in the WDPA either because they are not assigned or recorded by the Secretariats as part of their reporting protocols or because they have not been reported to the WDPA. This combination of missing and invalidated IUCN Categories undermines their utility for analysis. Therefore, analyses based on IUCN Categories can only provide a limited assessment of differences between the management objectives of protected areas.

In addition to gaps in attributes, missing or inaccurate protected area boundaries lead to the significant error in biodiversity protection statistics (Joppa & Pfaff, 2009; Visconti et al., 2013). Errors in spatial representation occur where point locations (assumed to be central coordinates) are the only descriptor available. In these cases, analysts must either exclude these sites altogether or try to use an approximation of area calculated through creating buffers around the known points. However, as protected area boundaries are very seldom symmetrical, estimating a boundary can be problematic in that all buffering methodologies differ in their geometric qualities and geographic parameters, creating pure circles or allowing for varying levels of distortion around the globe. Great strides have been made to obtain true boundaries for all sites, but little can be done at the global scale without improved capacity of governments to accurately survey and digitise their protected areas and then report them accurately and in a standardised format to the WDPA. By August 2014, 9% of the sites in the WDPA (19,548) were missing a boundary (IUCN and UNEP-WCMC, 2014).

Protected area statistics

Official estimates of protected area coverage based on the WDPA

The CBD established Aichi Target 11 in 2010 with the aim to protect 17% of the world's terrestrial and inland water areas and 10% of the world's marine areas by 2020 (CBD, 2010). Considerable progress has been made in expanding the protected area network, especially in the marine and coastal areas under national jurisdiction. While achieving the quantitative aspect of Aichi Biodiversity Target 11 seems on track, much more progress is needed to meet the qualitative aspects of the target, such as achieving effective and equitable protected areas, covering important places for biodiversity or developing well-connected protected area systems (Juffe-Bignoli et al., 2014a).

Terrestrial protected areas

By August 2014, protected areas recorded in the WDPA covered 15.4% of the world's terrestrial and inland water area, excluding Antarctica (Figure 5.3), equating to 20.6 million km^2 of the Earth, an area almost equivalent to the size of North and Central America combined (20.9 million km^2). The most extensive coverage achieved at a regional level has been for Central (28.2%) and South (23%) America (Juffe-Bignoli et al., 2014a). To meet the 17% terrestrial protection target of the CBD at a global level, an additional 2.2 million km^2 of terrestrial and inland water areas would have to be recognised as protected, an area roughly the size of Greenland (Juffe-Bignoli et al., 2014a).

Marine protected areas

By August 2014, marine protected areas (MPAs) recorded in the WDPA covered around 3.4% (12.3 million km^2) of the global ocean, with decreasing numbers of MPAs at greater distances from the shoreline. The highest proportion of marine protection occurs in territorial seas (0–12 nautical miles), where MPAs cover 10.9% of the nearshore waters. Protection drops to 5.4% in exclusive economic zones (0–200 nautical miles). To achieve 10% in all marine waters under national jurisdiction (AUNJ), a further 2.2 million km^2 of territorial seas and exclusive economic zones will need to be protected, an area six times the size of the Mediterranean Sea. In areas beyond national jurisdiction (ABNJ) (>200 nautical miles), MPAs are the result of multinational agreements; here, MPAs make up only 1% of the total ABNJ. To meet the 10% target for the global ocean, in addition to the area needed for AUNJ, a further 21.5 million km^2 of

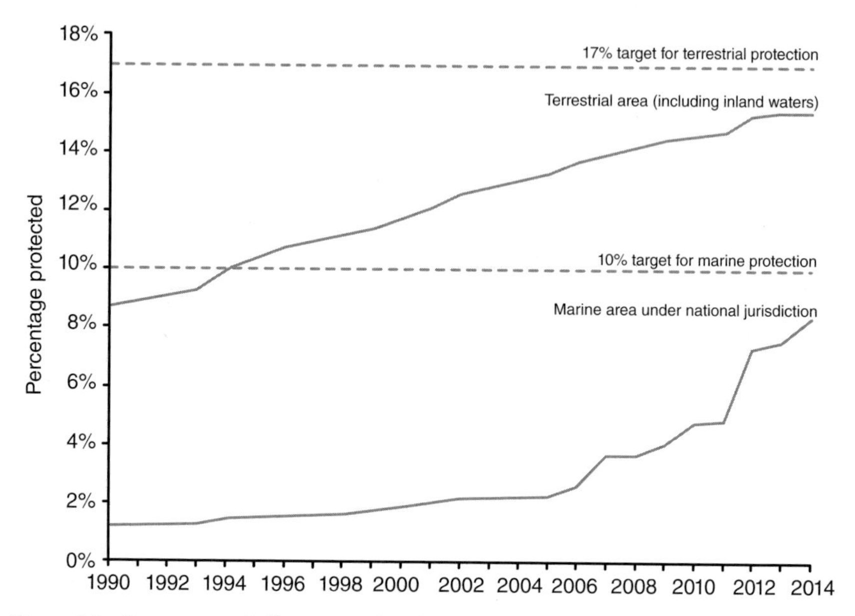

Figure 5.3 **Percentage of all terrestrial and marine areas (0–200 nautical miles) covered by protected areas, 1990–2014 (Juffe-Bignoli et al., 2014a)**

ABNJ would need to be protected. Figure 5.3 shows how these protection levels have changed over the last 24 years (Juffe-Bignoli et al., 2014a).

Biodiversity protection statistics

Targeting protected area coverage for biodiversity conservation

Aichi Target 11 requires protected area networks to 'be ecologically representative' and especially cover 'areas of particular importance for biodiversity and ecosystem services' (CBD, 2010). The CBD-mandated Biodiversity Indicators Partnership (BIP) has developed a number of indicators to assess the ecological representativeness of the global protected area network, as well as its coverage of the most important sites for biodiversity, including those supporting highly threatened species (BIP, 2010). The BIP indicators presented here measure:

- Ecological representativeness based on the world's terrestrial (Olson et al., 2001) and marine ecoregions (Spalding et al., 2007)
- Protected area coverage of two types of site-scale conservation priorities known as Key Biodiversity Areas (KBAs) (Eken et al., 2004):
  - Alliance for Zero Extinction sites (AZEs) (Ricketts et al., 2005)
  - Important Bird and Biodiversity Areas (IBAs) (Butchart et al., 2015)

Ecological representation

Ecoregions are large areas of land and sea with characteristic combinations of habitats, species and environmental conditions that are distinct from adjacent areas and vary by continent (Olson et al., 2001; Spalding et al., 2007). The previous CBD Strategic Plan and the 2002 World Summit on Sustainable Development first recognised ecological representativeness as a component of the then 10% protection target for terrestrial and marine areas globally. This component has carried over to the new CBD Aichi Targets and in particular Target 11. If the clause, 'ecologically representative', within Target 11 is taken literally, then the world should be aiming to achieve 17% protection of each terrestrial ecoregion and 10% of each marine ecoregion by 2020 (Woodley et al., 2012; Butchart et al., 2015). The 2014 Protected Planet Report showed that 43% of the 823 terrestrial ecoregions were protected at the 17% level and that 34% of the 232 marine ecoregions were protected at the 10% level (Juffe-Bignoli et al., 2014a).

While various analyses have been completed to show progress against politically agreed protected area targets, it should be noted that there is only a limited scientific basis for promoting 17%, 10% or other percentages of protection of all ecoregions to conserve biodiversity over the long term (Svancara et al., 2005; Noss et al., 2012; Locke, 2013; Butchart et al., 2015). More research is required to establish appropriate levels of protection for individual ecoregions, allowing for a more sophisticated interpretation of this element of the global protected area targets. Analyses of several of the focal biomes of the CBD (Box 5.1) provide a more detailed assessment. Based on the slow and unequal progress between 1990 and 2013, the task of expanding the global protected area network to cover 10% of all terrestrial and marine ecoregions by 2020, remains a significant challenge. Butchart et al. (2015) estimated that to achieve the 17% target for global and national terrestrial coverage and terrestrial ecoregions, an additional 10.5 million km^2 is needed.

Areas of particular importance for biodiversity

When tracking progress towards the 'areas of particular importance to biodiversity' aspect of Aichi Target 11 (CBD, 2010), KBAs stand out as the only sites of recognised conservation importance based on a globally standardised criteria (Eken et al., 2004; Langhammer et al., 2007; Butchart et al., 2015). AZEs (Ricketts et al., 2005) and IBAs (Grimmett & Jones, 1989) are the only two subsets of KBAs identified globally to date. AZEs alone are critical for the survival of one or more highly threatened species and therefore represent an urgent priority for protection (Ricketts et al., 2005). By 2013, only 20% (one-fifth) of AZEs and IBAs were covered by protected areas (Butchart et al., 2015). To protect all of the known important areas for biodiversity, in addition to the 17% target for global and national land coverage and ecoregions protection, Butchart et al. (2015) estimated that an additional 12 million km^2 of protected area would be needed.

Box 5.1 **Current protection statistics for CBD biomes (Juffe-Bignoli et al., 2014a)**

Dry and sub-humid lands: Dry and sub-humid lands make up the largest of the CBD biomes, covering approximately 40% of the world's land area and growing through desertification. Yet only 12.9% of this biome is protected.

Forests: Forests make up the second largest CBD biome, covering approximately 29% of the world's land area and shrinking. 20.1% of the world's forests are protected, with the highest levels of protection afforded through local communities and indigenous groups.

Inland Waters: This biome provides the most important ecosystem services to humans – water and food. 20.7% of all inland waters are protected with 7 out of 9 inland water types, defined by the Global Lakes and Wetlands Database, exceeding the 17% target.

Islands: Islands are home to 1/10 of the world's population and harbour large numbers of endemic species and unique ecosystems. Overall, 16.7% of the island biome is protected; however, temperate and polar islands individually have exceeded 17% protection. Protection is lowest in tropical islands (12.8%) where endemism is highest.

Mountains: Mountains cover approximately 25% of the world's land area outside Antarctica and are home to 22% of the world's population. Mountains occur in all major climate zones and are the source of half of the world's freshwater. 19.1% of the mountain biome is protected.

AZEs and IBAs alone provide a limited view of important areas for terrestrial biodiversity. Data sets for other types of KBAs, for example, freshwater taxa (Holland et al., 2012) and marine KBAs (so-called ecologically and biologically significant areas of the sea (Dunn et al., 2014)), are being developed and will provide a more comprehensive need for biodiversity protection.

Enabling the WDPA: fundamental principles

The WDPA is the baseline database for tracking progress towards globally agreed protected area targets. Yet the WDPA's utility depends on three key factors:

1. A globally agreed definition and/or criteria for what areas count towards biodiversity protection
2. National to global data sharing
3. Adequate resourcing

In many ways, it would seem that the prerequisites for enabling the WDPA to fulfil its mandates accurately have been achieved but there are significant pitfalls.

What counts as a biodiversity protection area

Many governments have not fully recognised the extent of their own protected area networks, reporting only statutory protected areas at federal and sub-national levels of governance. However, the IUCN and CBD recognise a wider array of governance types, including ICCAs and PPAs (Dudley, 2008; CBD, 2014). The causes for the omission of these areas on national lists are varied, but experience liaising with countries to update the WDPA has shown that the root causes are a lack of political priority to compile non-governmental protected areas into national lists and a disagreement between the international definition of a protected area and autonomous national definitions. The only way many non-statutory protected areas make it into the WDPA is through non-governmental data providers. The total non-statutory protected area estate represented in the WDPA as being governed by indigenous peoples and local communities, by private entities (including NGOs) or under shared governance had increased substantially from about 2.4% in 1990 to 10.9% in 2014 (Juffe-Bignoli et al., 2014b). However, new studies and reports from the conservation community are showing that the full contribution of these areas is far from realised (Carter et al., 2008; Roe et al., 2009; Borrini-Feyerabend et al., 2010; Verschuuren et al., 2010; Kothari, 2012, 2014; Porter-Bolland et al., 2012; Borrini-Feyerabend, 2014; Stolton, 2014).

ICCAs

The contribution of ICCAs to biodiversity conservation, sustainable development and human rights has gained a great deal of attention (Verschuuren et al., 2010; Porter-Bolland et al., 2012; Borrini-Feyerabend et al., 2014; Kothari et al., 2014). However, the number of protected areas governed by indigenous peoples and local communities still only amounted to 1% of the WDPA by 2014 (Juffe-Bignoli et al., 2014b). In contrast, the total area of ICCAs contributed to almost 5% of the total protected area estate as represented by the WDPA (Juffe-Bignoli et al., 2014b). Some estimates suggest their total area may cover as much as 13% of the Earth's landmass (Kothari et al., 2012). Community forests, for example, exist on every inhabitable continent and may contribute at least as great an area conserved by state governments (Kothari et al., 2014). This particular type of ICCA is also rapidly expanding through efforts to promote community-based forest and wildlife management approaches, which are particularly popular in Africa (Roe et al., 2009). In the marine waters surrounding Japan, Spain, parts of Africa (notably Madagascar and Kenya), southeast Asia and the south Pacific, locally managed marine areas are governed by local communities to protect coastal and marine resources – the greatest of these representing an estimated 20,000 km^2 in the south Pacific (Kothari, 2014).

In addition to the WDPA, UNEP-WCMC together with UNDP, the ICCA Consortium and a wide range of partners has developed the ICCA Registry (www.iccaregistry.org), a database that currently includes information and case studies on ICCAs. However, assessing the extent of ICCAs and incorporating them in the WDPA are complicated by the fact that many ICCAs do not have clearly defined boundaries but exist within the boundaries of the indigenous or community-owned lands. Some countries are indirectly addressing this issue through new national legislation that seeks to recognise the conservation status of such lands, thus making them more easily accountable. For example, Australia has developed a category of Indigenous Protected Areas (IPAs) within its national reserve system where communities are able to go through a consultation process and decide whether or not they want to become officially declared IPAs (Commonwealth of Australia, 2011).

PPAs

In addition to statutory protected areas and ICCAs, many protected areas are managed for the long-term conservation of biodiversity on land that is privately owned. Although PPAs have been around for centuries (Carter et al., 2008; Stolton et al., 2014), they have largely been neglected by the conservation community due to lack of available data, policy guidance and understanding of their growth and contribution to biodiversity protection (Stolton et al., 2014). The lack of clear national and global definitions for PPAs and guidance as to what IUCN or CBD considered a PPA is also a problem (Stolton et al., 2014). The gaps in accounting for PPAs are at least part of the reason governments seldom included them in national lists and therefore are not reported to the WDPA. Very few countries seem to have any initiatives in place to collect data on PPAs at a national scale. One exception is the United States, where the US Geological Society's Gap Analysis Program collects information on land trusts, easements and other PPAs as part of its Protected Areas Database (PADUS) (USGS, 2012).

In October 2013, UNEP-WCMC estimated a possible 17,505 PPAs recorded in the WDPA based on a combination of records with private governance types allocated and records with names or designations including the word 'private' (Stolton et al., 2014). However, the Protected Planet Report 2014, based on the more recent August 2014 version of the WDPA, reported only 9500 records of PPAs based on governance type alone (Juffe-Bignoli et al., 2014b), almost half what it had reported a year earlier. Known PPA gaps in the WDPA include several African countries, where private conservancies often constitute large-scale protected areas, often under complex tenure arrangements (Carter et al., 2008). In Australia, conservation covenants and private land trusts could add almost 5,000 PPAs to the protected area network covering 89,130 km^2 of land (Fitzsimmons, 2015).

A challenge for defining and including PPAs with IUCN and WDPA standards will be to assess their permanency in accordance with the IUCN definition, which states an area must be protected in perpetuity as one aspect of achieving protected area status

(Dudley, 2008). Many PPAs are not intended to surpass the life of the ownership entity, be it an individual, company or NGO or may have specified time limits on their existence, such as mitigation areas protected by companies for the life of an operation.

Other effective area-based conservation measures

The 'other effective area-based conservation measures' clause of Aichi Target 11 (CBD, 2010) seems to extend biodiversity protection beyond the known protected area network, but just how far is up for debate. There is no definition or guidance as of yet as to what types of designated land uses may qualify as OECMs (Lopoukhine & de Souza Dias, 2012; Woodley et al., 2012; UNEP-WCMC, 2015). However, current arguments caution that the intent of OECMs should match that of protected areas (Woodley et al., 2012) and emphasise long-term biodiversity conservation as key criteria stating that OECMs may simply refer to non-statutory protected areas that are under-represented in the WDPA, that is, PPAs and ICCAs (Lopoukhine & de Souza Dias, 2012; Woodley et al., 2012).

National to global data sharing

The United Nations List of Protected Areas and the national reporting obligations of CBD, such as National Biodiversity Strategy and Action Plans, have helped to facilitate national to global data sharing. However, these reports and data flows are not based on a unified understanding and interpretation of what counts towards biodiversity protection, and compilation of national reports is still based on methods of manual documentation rather than through seamless, semi-automated data management systems. Providing data to the WDPA still requires a significant parallel effort by countries. Likewise, keeping the WDPA up to date requires substantial time and resources from UNEP-WCMC, NGOs, consortiums and expert individuals. Further complicating matters, some data providers restrict the level of access to their protected areas data. Some are reluctant to disseminate their data publically because they would prefer to vet and keep track of each request to use the data. Others are guarding against the potential for commercial entities to use their data for monetary gain or are guarding against the illegal taking of resources, such as poaching. Other reasons cited for restricting public dissemination of data are to prevent the loss of version control, loss of intellectual property or the desire for monetary gain from the sale of data though these are becoming less common.

Adequate resourcing

The third pillar supporting the WDPA is resource allocation. Throughout its history, the WDPA has survived funding hardships. However, every lean financial period created gaps in progress and losses in data quality, which undermined confidence in the WDPA as a viable product, and eventually led to a costly recovery period. In the last two decades, the WDPA has gained financial support primarily through the Proteus Partnership, a unique collaboration between leading extractives companies and the UNEP-WCMC. Additional support has come from the UNEP Challenge Fund, IUCN WCPA and various projects. However, the lack of a core-funding model makes consistent database development a difficult intention to keep as development is driven by the aims of the projects or the priorities of external partners. Although the aim of the Proteus Partnership for the WDPA has been simply to deliver a quality data set (UNEP-WCMC, 2012), it is not a guaranteed future funding model, and there is a huge gap in project funding specifically for data acquisition, database development, content management or even analysis.

Data stewardship is a solution

Divesting some of the WDPA maintenance to data providers and making better use of technology are promising solutions to data sharing and resource constraints on the WDPA. In this way, data providers would become stewards of their own data in the WDPA. WDPA stewardship development could involve upload tools for data providers or dynamic links between national databases and the WDPA, replacing as much manual intervention as possible. This kind of stewardship could potentially speed up the updates process to an almost automated and real-time solution. Dynamic data stewardship is tantalising and has been tested to some degree through web services and cloud hosting, but the technology does not yet exist fully to support updates to the WDPA without requiring a significant amount of human validation and integration procedure.

Updates to the WDPA come from hundreds of sources where data is stored in as many different formats to fulfil an equal variety of local purposes. Assuming every CBD member state in the world would participate in dynamic data stewardship, over 100 customised pathways would need to be developed. The most difficult requirements are those that cannot be overcome by technology alone; they are requirements for open and unrestricted access to data, ability and willingness to use the appropriate technology on both ends of the data flow and a shared understanding of what should be included in the updates. To highlight the magnitude of the challenge, several developing countries still do not have their own official GIS data sets.

Apart from automatic updates, there is a simpler, cost-effective and immediate way to improve data stewardship in the WDPA. The first critical step is adoption of the WDPA Data Standard and in particular inclusion of the unique WDPA Identifier (WDPA ID) in the source data sets. Once the WDPA ID is included in or mapped to source data sets, updates can occur much more efficiently at the hands of the WDPA team; more importantly, adoption of the WDPA ID is the first, critical step towards automated updates. The WDPA ID is already included as an attribute in several global, regional and national protected area databases. The most significant of these by sheer proportion of sites is the European Environment Agency's Common Database on Designated Areas (CDDA). Owing to the similar structures of the two databases, the robust validation procedures in place by EEA and inclusion of WDPA ID into the CDDA Site Code, updates for 37 countries in Europe – one-third of the WDPA – can occur on an annual basis in less time than it takes to update some individual country data sets. Other initiatives that rely on the continuity of the WDPA ID include the GDPAME, Integrated Biodiversity Assessment Tool (IBAT), the DOPA, WWF's PADDD tracker and ArkGIS platforms and NASA's Fire Information and Resource Management System (FIRMS).

Summary and conclusion

The WDPA is used to answer questions about how well we are protecting biodiversity and the life-sustaining resources of our planet. As the official protected area database of UNEP and IUCN, the WDPA is the only authoritative, global database on protected areas and may soon have to change its name as clarity on the full gamut of areas benefiting conservation are incorporated into its remit. The WDPA has come a long way over the last five decades to receive recognition as a key tool to assist in the monitoring of progress towards achieving protected area targets. With its UN List mandate, extensive policy decisions successful data standard and global user base, the WDPA is well positioned to continue its evolution and purpose long into the 21st century.

The WDPA is continually evolving as evidenced by updates to the WDPA Data Standard in 2015 and the publication of the WDPA Manual (UNEP-WCMC, 2015). Quality is achieved by addressing the conditions for which data is submitted to the WDPA including a minimum requirement for key attributes on which analyses of biodiversity protection depend. Closing gaps within the WDPA will remove a significant margin of known error in the database, but it will be even more important to ensure all areas that count towards biodiversity protection, such as ICCAs and PPAs, are included. Enhanced internal and external validation procedures and expert reviews, both of which are a significant part of a new WDPA quality assurance process, will help to assess errors and omissions more efficiently.

The WDPA shows that protected areas have increased dramatically over the last century. Nations of the world appear to be on a good trajectory towards meeting the biodiversity coverage requirements of Aichi Target 11. However, this target challenges the world to look beyond extent of protected areas alone, requiring that the global protected area network is supported by equitable governance and effective management and is ecologically representative and well connected. We now have the knowledge to ensure that new protected areas are placed in areas where they will have the greatest potential to protect biodiversity. An increasing number of global studies using the WDPA show that protected areas, as a conservation strategy, are indeed making a critical contribution to habitat and species conservation (Geldmann et al., 2013; Venter et al., 2014; Butchart et al., 2015).

As the foundation stone for measuring biodiversity protection, it is critical to understand the factors on which the WDPA depends for success as these can also undermine its utility. Global agreement on what counts towards biodiversity protection, unrestricted national to global data sharing and adequate resourcing are key to ensuring the continued viability of the WDPA. In order for IUCN and UNEP to ensure the WDPA is fit for purpose to track progress towards all components of Aichi Target 11 and other targets relevant to protected areas and biodiversity protection, they will have to maintain development of the WDPA in parallel with the development of indicators and data. Furthermore, as rapid improvements in technology are realised disproportionately across the world, WDPA accuracy will become an ever more challenging but crucially important part of its development. The greatest potential to maintain WDPA relevance and accuracy into the future is through investment in the WDPA, data stewardship, dynamic linkages with other biodiversity-relevant databases, renewed partnerships and the backing of the conservation community at large.

Note

1. IUCN definition: 'A protected area is a clearly defined geographical space, recognised, dedicated and managed, through legal or other effective means, to achieve the long term conservation of nature with associated ecosystem services and cultural values' (Dudley, 2008). CBD definition: 'a geographically defined area, which is designated or regulated and managed to achieve specific conservation objectives' (UN, 1992).

References

Anderson, S., 2002. *Identifying important plant areas*, London: Plantlife International.

Bertzky, B. et al., 2012. *Protected planet report 2012: Tracking progress towards global targets for protected areas*, Gland & Cambridge: IUCN & UNEP-WCMC.

Besancon, C., 2007. Conservation data for smarter business decisions. Business 2010: Technology Transfer and Cooperation under the Convention, September 2007, 2(3), p. 16–17.

BIP, 2010. *Biodiversity indicators and the 2010 target: Experiences and lessons learnt from the 2010 biodiversity indicators partnership*, Montreal: Secretariat of the CBD.

Borrini-Feyerabend, G. et al., 2010. *Biocultural diversity conserved by indigenous peoples and local communities: Examples and analysis*, Tehran: CUED.

Borrini-Feyerabend, G. P. B. et al., 2014. A primer on governance for protected and conserved areas, Stream on Enhancing Diversity and Quality of Governance. 2014 IUCN World Parks Congress. Gland, IUCN.

Butchart, S. et al., 2015. Shortfalls and solutions for meeting national and global conservation area targets. *Conservation Letters*, p. 1–9.

Carter, E., Adams, W. M. & Hutton, J., 2008. Private protected areas: Management regimes, tenure arrangements and protected area categorization in East Africa. *Oryx*, 42(2), p. 177–186.

CBD, 2010. *Decisions adopted by the CoP to the CBD at its 10th meeting (UNEP/CBD/COP/DEC/X/2)*, Montreal: Secretariat of the CBD.

CBD, 2014. *Decisions adopted by the CoP to the CBD at its 12th meeting (UNEP/CBD/COP/DEC/XII/19)*, Montreal: Secretariat of the CBD.

Coad, L. et al., 2013. Progress towards the CBD protected area management effectiveness targets'. *PARKS*, 19(1), p. 13–24.

Commonwealth of Australia, 2011. Indigenous Protected Areas. [Online] Available at: http://www.environment.gov.au/indigenous/ipa/index.html [Accessed 7 September 2011].

Deguignet M. et al., 2014. *2014 United Nations list of protected areas*, Cambridge: UNEP-WCMC.

Dubois, G. et al., 2015. *The digital observatory for protected areas (DOPA) explorer 1.0*, Ispra: European Commission, Joint Research Centre.

Dudley, N., 2008. *Guidelines for applying protected area management categories*, Gland: IUCN.

Dunn, D. et al., 2014. The convention on biological diversity's ecologically or biologically significant areas: Origins, development, and current status. *Marine Policy*, 49, p. 137–145.

Eken, G. et al., 2004. Key biodiversity areas as site conservation targets. *BioScience*, 54(12), p. 1110–1118.

Environment Canada, 2015. Canada's Protected Areas. [Online] Available at: http://www.ec.gc.ca/indicateurs-indicators/default.asp?lang=en&n=478A1D3D-1 [Accessed 13 June 2015].

Fitzsimmons, J., 2015. Private protected areas in Australia: Current status and future directions. *Nature Conservation*, 10, p. 1–23.

Fuller, R. A. et al., 2010. Replacing underperforming protected areas achieves better conservation outcomes. *Nature*, 466, p. 365–367.

Geldmann, J. et al., 2013. Effectiveness of terrestrial protected areas in reducing habitat loss and population declines. *Biological Conservation*, 161, p. 230–238.

Grimmett, R. F. A. & Jones, T. A., 1989. *Important bird areas in Europe*, Cambridge: Birdlife International.

Holland, R. A., Darwall, W. R. T. & Smith, K. G., 2012. Conservation priorities for freshwater biodiversity: The key biodiversity area approach refined and tested for continental Africa. *Biological Conservation*, 148(1), p. 167–179.

IUCN and UNEP-WCMC, 2014. World Database on Protected Areas. [Online]. August 2014. Cambridge: UNEP-WCMC. Available at: www.protectedplanet.net [Accessed 10 July 2015].

Jenkins C. N. & Joppa, L. N., 2010. Considering protected area category in conservation analysis. *Biological Conservation*, 143(1), p. 7–8.

Jenkins, C., Pimm, S. & Joppa, L. N., 2013. Global patterns of terrestrial vertebrate diversity and conservation. *Proceedings of the National Academy of Sciences of the United States of America*, 110(28), p. E2602–E2610.

Jonas, H. et al., 2014. New steps of change: Looking beyond protected areas to consider other effective area based conservation measures. *PARKS*, 20.2, 111–128.

Joppa, L. & Pfaff, A., 2009. High and far: Biases in the location of protected areas. *PLoS ONE*, 4(12), p. e8273.

Juffe-Bignoli, D. et al., 2014a. *Asia protected planet 2014*, Cambridge: UNEP-WCMC.

Juffe-Bignoli, D. et al., 2014b. *Protected planet report 2014*, Cambridge: UNEP-WCMC.

Kingston, N. et al., 2015. Knowledge generation, acquisition. In: G. Worboys, M. Lockwood & A. Kothari, eds. *Protected area governance and management*. Canberra: ANU Press, p. 327–352.

Knowles, J. et al., 2015. Establishing a marine conservation baseline for the insular Caribbean. *Marine Policy*, 60, p. 84–97.

Kothari, A. et al., 2012. Recognising and Supporting Territories and Areas Conserved by Indigenous Peoples and Local Communities: Global Overview and National Case Studies. Technical series no. 64., Montreal: Secretariat of the Convention on Biological Diversity, ICCA Consortium, Kalpavriksh, and Natural Justice.

Kothari, A. et al., 2014. ICCAs and Aichi Targets: The Contribution of Indigenous Peoples' and Local Community Conserved Territories and Areas to the Strategic Plan for Biodiversity 2011-20. Policy Brief of the ICCA Consortium, s.l.: CBD Alliance, Kalpavriksh, CENESTA and IUCN Global Protected Areas Programme.

Langhammer, P. F. et al., 2007. *Identification and gap analysis of key biodiversity areas: Targets for comprehensive protected area systems*, Gland: IUCN.

Leadly, E. A., 2014. Progress toward the Aichi Biodiversity Targets: An assessment of biodiversity trends, policy scenarios and key actions. Technical Series 78, Montreal: Secretariat of the CBD.

Locke, H., 2013. Nature needs half: A necessary and hopeful new agenda for protected areas. *PARKS*, 19, p. 9–18.

Lopoukhine, N. & de Souza Dias, B. F., 2012. Editorial: What does target 11 really mean. *PARKS*, 18(1), p. 5–9.

Mascia, M. B. & Pailler, S., 2011. Protected area downgrading, downsizing and degazettement (PADDD) and its conservation implications. *Conservation Letters*, 4, p. 9–20.

Mascia, M. et al., 2014. Protected area downgrading, downsizing, and degazettement (PADDD) in Africa, Asia, and Latin America and the Caribbean, 1900–2010. *Biological Conservation*, 169, p. 355–361.

Molnar, A., Scherr, S. & Khare, A., 2004. *Who conserves the world's forests? Community driven strategies to protect forests and respect rights*, Washington, DC: Forest Trends.

Noss, R. F. et al., 2012. Bolder thinking for conservation. *Conservation Biology*, 26(1), p. 1–4.

Olson, D. M. et al., 2001. Terrestrial ecoregions of the world: A new map of life on earth. *BioScience*, 51(11), p. 933–938.

Porter-Bolland, L. et al., 2012. Community managed forests and forest protected areas: An assessment of their conservation effectiveness across the tropics. *Forest Ecology and Management*, 268, p. 6–17.

Ricketts, T. H. et al., 2005. Pinpointing and preventing imminent extinctions. *Proceedings of the National Academy of Sciences of the United States of America*, 102(51), p. 18497–18501.

Roe, D., Nelson, F. & Sandbrook, C., 2009. *Community management of natural resources in Africa: Impacts, experiences and future direction*, Natural Resource Issues No. 18, London: IIED.

Spalding, M. D. et al., 2007. Marine ecoregions of the world: A bioregionalization of coastal and shelf areas. *BioScience*, 57(7), p. 573–583.

Spalding, M. et al., 2013. Protecting marine spaces: Global targets and changing approaches. In: A. Chircap, S. Coffen-Smout & M. McConnell, eds. *Ocean yearbook 27*. Boston, MA: Martinus Nijhoff, p. 213–248.

Stolton, S. R. K. et al., 2014. *The futures of privately protected areas*, Gland: IUCN.

Svancara, L. K. et al., 2005. Policy driven versus evidence based conservation: A review of political targets and biological needs. *BioScience*, 55(11), p. 989–995.

UN, 1992. *Convention on biological diversity*, New York: UN.

UNEP, 2014. *The millennium development goals report 2014*, New York: UNEP.

UNEP-WCMC, 2012. *Rio+20 Briefing Note No. 4: Integrating biodiversity and ecosystem services into business decision-making*, Cambridge: UNEP-WCMC.

UNEP-WCMC, 2015. *The world database on protected areas user manual 1.0*, Cambridge: UNEP-WCMC.

USGS, 2012. *Protected areas database of the United States: Standards and methods manual for data stewards*, Boise: USGS Gap Analysis Program at Boise State University.

Venter O. et al., 2014. Targeting global protected area expansion for imperiled biodiversity. *PLoS ONE*, 12(6), p. e1001891.

Verschuuren, B., Wild, R., McNeely, J. & Oviedo, G., 2010. *Sacred natural sites: Conserving nature and culture*, London: Earthscan.

Visconti, P., Di Marco, M., Alvarez-Romero, J. G. & Januchowski-Hartley, S. R., 2013. Effects of errors and gaps in spatial data sets on assessment of conservation progress. *Conservation Biology*, 27(5), p. 1–11.

Watson, J. E. M. et al., 2010. The capacity of Australia's protected area system to represent threatened species. *Conservation Biology*, 25, p. 324–332.

WDPA, 2010. *The World Database on Protected Areas (WDPA)*, Gland & Cambridge: IUCN & UNEP-WCMC.

WDPA, 2014. *The World Database on Protected Areas (WDPA)*, Gland & Cambridge: IUCN & UNEP-WCMC.

White, A. & Martin, A., 2002. *Who owns the world's forests? Forest tenure and public forests in transition*, Washington, DC: Forest Trends & Centre for International Environmental Law.

Woodley, S. et al., 2012. Meeting Aichi target 11: What does success look like for protected area systems. *PARKS*, 18(1), p. 23–36.

Part II
The Fate of Species in Protected Areas

6

Species Population Trends in Protected Areas

B. Collen[1], *L. McRae*[2], *E. Nicholson*[3,4], *I. D. Craigie*[5], *E. J. Milner-Gulland*[6,7], *J. Loh*[8,9] *and S. Whitmee*[1,2,10]

[1]Centre for Biodiversity & Environment Research, University College London, London, UK
[2]Institute of Zoology, Zoological Society of London, London, UK
[3]School of Botany, University of Melbourne, Melbourne, Victoria, Australia
[4]Deakin University, Geelong, Australia. School of Life and Environmental Sciences, Centre for Integrative Ecology (Burwood Campus), Australia
[5]ARC Centre of Excellence for Coral Reef Studies, James Cook University, Townsville, Queensland, Australia
[6]Department of Life Sciences, Imperial College London, Berkshire, UK
[7]Department of Zoology, Oxford University, Oxford, UK
[8]WWF International, Gland, Switzerland
[9]School of Anthropology and Conservation, University of Kent, Canterbury, UK
[10]International Union for Conservation of Nature, Cambridge, UK

Introduction

Land-use change and unsustainable exploitation threaten large numbers of species worldwide. Protected areas are increasingly becoming the last refuges for many wild species (Butchart et al., 2012; McKee et al., 2013). However, despite their importance in maintaining viable populations of wildlife and protecting critical habitats over the long term (Newman, 2008), there is no agreed system for monitoring the state of wildlife in and around these protected areas (Chape et al.,

Protected Areas: Are They Safeguarding Biodiversity?, First Edition.
Edited by Lucas N. Joppa, Jonathan E. M. Baillie and John G. Robinson.

2005). If informed decisions are to be made about the management of wildlife populations in protected areas, populations need to be monitored (Ferraro & Pattanayak, 2006). Given that the decline in populations is a prelude to species extinction (Ceballos & Ehrlich, 2002; Collen et al., 2011) and tends to reduce taxonomic, genetic, and functional diversity, abundance is recognized as a key measure of ecosystem health, which is integrally related to many different elements of biodiversity (Mace et al., 2008). By tracking trends in species abundance, that is, change in the number of individuals over time, it is possible to assess how well protected areas are meeting one of their primary functions, namely, to sustain and increase the abundance of the species they contain (Collen et al., 2009).

Measuring rate of change of population size is, perhaps, one of the most sensitive metrics for long-term measurement of biodiversity change (Balmford et al., 2003; Buckland et al., 2005; Pereira & Cooper, 2006). It is a trend that can be updated annually, which is important given the immediacy with which biodiversity information is needed. Abundance measures are an important proxy for biodiversity at higher levels and can be used to infer community change (Buckland et al., 2005). Change in abundance also indicates the effects of processes such as exploitation/overharvesting and habitat loss and therefore provides an early warning system for the loss of essential ecosystem services and harvested goods (Balmford et al., 2003).

The aggregation of monitored trends in animal population abundance is a key tool with which to track the changing fate of wildlife at global (Gregory et al., 2005; Loh et al., 2005; Collen et al., 2009; McRae et al., 2012) and national levels (McRae et al., 2007; Pomeroy & Tushabe, 2010; Collen et al., 2013). One technique developed to monitor broadscale changes in abundance, the Living Planet Index (Loh et al., 2005), has recently been applied to assess trends in large mammals in African protected areas. Results indicate that African protected areas have generally failed to mitigate human-induced threats to large mammal populations (Craigie et al., 2010), a result recently echoed for African bird species (Scholte, 2011; Beale et al., 2013). These kinds of analyses rely on large quantities of baseline monitoring data across a broad set of species, ideally with appropriate counterfactual data for populations located outside protected areas (Western et al., 2009; Chessman, 2013).

Despite the broad application of analyses that seek to evaluate trends in wildlife abundance in and around protected areas, they have been beset with difficulties. Most frequently, there is a lack of available data with which to evaluate the efficacy of protected areas in protecting wildlife populations. To date, monitoring has tended to be undertaken in a haphazard, non-systematic manner in the majority of locations. In this chapter, we ask whether or not population trend data can provide a useful basis, from which a greater understanding of the role, efficacy, and coverage of protection can be established. We build on previous work on indicators of abundance trends to review three of the most important areas of research pertaining to trends of wildlife in protected areas. Specifically, we evaluate:

- Data availability: We review what monitoring data are available at a global scale to answer questions concerning species population change in and around protected areas over time.
- Current trends: We present a case study of how species population trends have been evaluated to measure broadscale trends in African protected areas.
- Future trends: We evaluate how abundance trends can be used in a more effective manner to support decision making about protected area policy through scenario modeling.

For the first section, we explore the trends in species populations of vertebrate species in protected areas between 1950 and the present day contained in the Living Planet Index (2012) database. This resource provides, to our knowledge, the largest compendium of abundance trend data for vertebrates, but has hitherto not been used to evaluate trends in protected areas at the global scale. We use this dataset to evaluate trends, identify gaps in abundance monitoring, and speculate what would be required to implement a monitoring system to discern between the effectiveness of protected areas and areas beyond their borders and describe what such a system might look like.

For the second section, we provide a practical demonstration of how available trend data can be used for monitoring broadscale trends in abundance by examining an example of how trends in mammal abundance in African protected areas might be used to determine broadscale trends.

In the third section, we examine how population abundance trend information might be used to inform protected area policy by asking whether the types of indicators designed to inform global, regional, and national biodiversity trends can be used to discern between policy options in protected area management. Using an adaptive management framework, we evaluate how scenario modeling might be used to estimate differences in outcome between several hypothesized management scenarios.

Data availability: Collating abundance data for protected area evaluation

The first task in tracking global trends in the changing fate of wildlife populations in protected areas is to compile existing monitoring data. The data collection protocols we outline here have been extensively described elsewhere (Loh et al., 2005; Collen et al., 2009). Briefly, data on vertebrate abundance trends over time were collated from published scientific literature, online databases (NERC Centre for Population Biology, 2010; BirdLife International, 2011), and unpublished nongovernmental organization and national park records literature. Data were only included if the following set of criteria were met (Collen et al., 2009):

- A measure of population size was available for a minimum of 2 years.
- The data were collected using the same method on the same population throughout the time series.
- Information was available on how the data were collected and what the units of measurement were.
- The geographic location of the population was provided.
- Data source was referenced and traceable.

All abundance trend data are available through the Living Planet Index (2012) database. Data on protected area extent and location were downloaded from the World Database on Protected Areas (WDPA, 2010). Abundance time series data were matched to the global WDPA dataset using study location data or direct references from within time series source material.

The abundance trend dataset consisted of 4337 populations of 1543 species in 977 unique protected areas listed in the WDPA (some data are from protected areas with no WDPA designation). Terrestrial and freshwater systems show higher data availability (3212 abundance time series for 1105 species from 643 protected areas) compared to marine systems (1125 abundance time series for 484 species from 334 protected areas) (Figure 6.1a). Note that a small number of species may have populations in both marine and terrestrial/freshwater systems; therefore, totals for marine and land do not sum to the total species given above. The quantity and variety of time series data are depicted in Figure 6.1b–d. The number of time series data shows a skewed distribution over the time period 1950–2012 (Figure 6.1b). A rapid increase in monitoring programs is apparent since 1970, peaking in 1997 when over 2308 populations had multiyear monitoring programs in place, before a rapid decline back to near zero toward the present, due to a lag between the recording of data and its publication. A plot of time series length (Figure 6.1c) shows that monitoring programs in protected areas range from 2 to 60 years in length (mean = 13.78 years, median = 11 years). Figure 6.1d shows the incidence of new monitoring schemes over time (upper panel) and also those for which data collection on abundance trends ends (lower panel). This last figure raises some questions about the longevity of monitoring programs. Specifically, it shows that a high number of monitoring programs were being implemented particularly in the 1990s. However, it also shows a high number of monitoring programs apparently ending and therefore unable to contribute to the understanding of trends in vertebrate abundance in protected areas.

Abundance data were available for protected areas in 133 countries worldwide. However, of the 130,709 protected areas defined in the WDPA (2010), only 997 had vertebrate abundance trend records for at least one species, with the number of available population time series in each protected area ranging widely from 1 to 102 species (mean = 3.74 species, median = 1 species). Protected sites with high

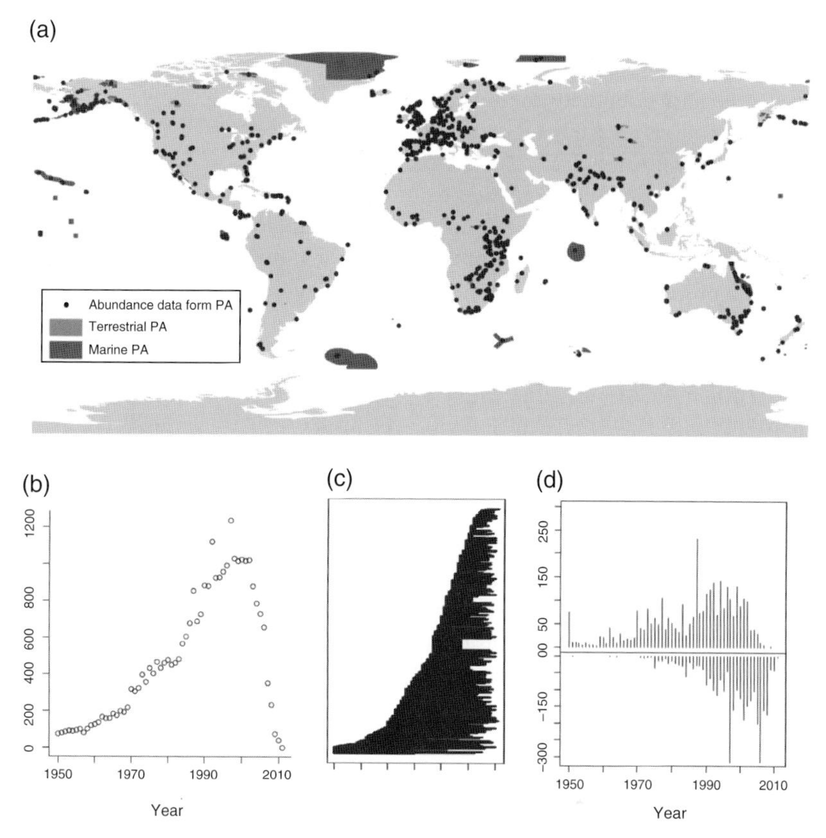

Figure 6.1 **Location of abundance time series in protected areas. (a) Map showing coincidence of marine and land-based protected areas with population abundance data. (b) Number of time series data over time. (c) Length of time series data. (d) Time series starting and ending. Source: Data from www.livingplanetindex.org. © Zoological Society of London and WWF, 2014**

data availability included: Doñana National Park in Spain (102 species), Serengeti National Park in Tanzania (21 species), and the Great Barrier Reef in Australia (11 species). Data from existing monitoring schemes were lacking from protected areas in South and Central America and Southeast Asia and from pelagic marine protected areas (Figure 6.1a).

Current trends: African case study

Time series data (Figure 6.1a–d) have a range of uses to aid management and provide information about current trends in wildlife in protected areas. In this section, we illustrate one way in which this has been achieved by aggregating time series into an index of change in abundance for vertebrates in African protected areas, an approach previously described by Craigie et al. (2010). This dataset has a high degree of overlap with the Living Planet Index database; however, a limited number of time series are not redistributed, due to restrictions from the holders of the original monitoring data.

The detail of the index calculation technique is fully reported (Loh et al., 2005) and was subsequently updated to a generalized additive modeling framework (GAM) (Collen et al., 2009). To produce an index, population abundance time series data of greater than 5 years in length were aggregated to calculate average rates of change in populations of vertebrate species (Craigie et al., 2010). This results in an index where values relate to the proportional change in population size over time, integrating the net impacts of populations of species both gaining and declining in number, to give an average proportional change in abundance between successive years.

Two complementary methods were used to generate index values, depending on time series length: a chain method (Loh et al., 2005) and a GAM method (Collen et al., 2009). While there are potential issues for index calculation resulting from chaining (Frisch, 1936; Ehemann, 2007), GAMs are not possible on datasets with small numbers of population estimates (Fewster et al., 2000; Buckland et al., 2005). Choice of method was therefore determined on the length of the time series, with times series of $n > 6$ data points being processed using the GAM framework, and those that did not meet the criteria processed using the chain method. In order to calculate an index, the logarithm of the ratio of population measure for each species was calculated for successive years. Mean values were calculated for species with more than one population. The overall index was then calculated with the index value set to 1 in 1970. Insufficient population time series are currently available to continue the index beyond 2007 because of a lag in the publication of data. Indices are produced weighting populations equally within species, and species weighted equally within each index. The index value presented is then a proportional measure of change in abundance relative to a value of 1.0 at the start of the index, with a value of 0.5 equating to a 50% decline and a value of 1.5 equating to a 50% increase in average population abundance of the populations contributing to the index.

Time series data were available for 583 mammal populations of 69 species from 78 African protected areas (Craigie et al., 2010). The data suggest that monitored African protected areas sustained a decline in the size of their large mammal populations between 1970 and 2005, with an overall index value decreasing from 1.00 to 0.41 (95% CI 0.34–0.50) (Figure 6.2a), indicating on average a 59% decline in abundance. Stark regional differences were apparent, with the east African regional index showing a similar rate of change to that of the continent-wide index (Figure 6.2b), western Africa showing the largest regional decline in abundance to an index value of around 0.15 (95% CI 0.10–0.22) (Figure 6.2c), and southern Africa showing a positive trend with an index value of 1.24 in 2005 (95%

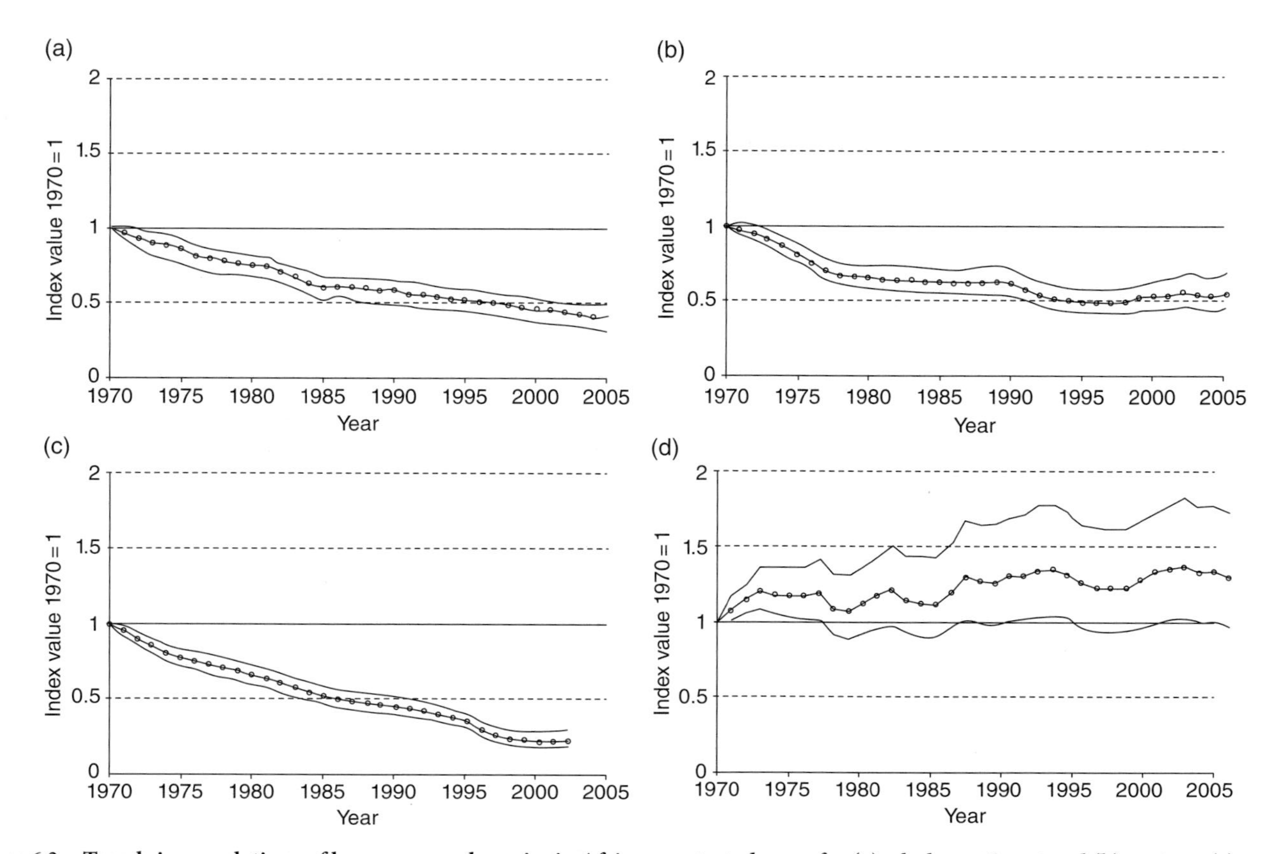

Figure 6.2 **Trends in populations of large mammal species in African protected areas for (a) whole continent and (b) eastern, (c) western, and (d) southern Africa. Source: From Craigie et al. (2010). Reproduced with permission of Elsevier**

CI 0.89–1.72) (Figure 6.2d). Wide confidence limits were apparent in particular for the southern Africa index, indicating a heterogeneous response of monitored populations within protected areas.

Future trends: Scenario models to support protected area policy

Predicting the plausible future trajectory of trends of wildlife in protected areas has rarely been attempted but is a key step toward making more informed decisions about the wildlife that they aim to protect. Modeling the impact of future scenarios of management choices or environmental condition gives the option of comparing different, or even competing, management interventions or policies. The concept behind the scenario models displayed in this chapter was devised by Nicholson et al. (2012) and is fully detailed in Costelloe et al. (2015). Six hypothetical scenarios of future management were compared, in order to evaluate how current trends in vertebrate abundance would fare under different continental-scale policies for African terrestrial protected areas, in comparison to a business as usual scenario. The aim of the study was not to provide a detailed analysis of the effects of these policies on the abundance of mammals in and around protected areas. Rather, it was to assess whether the indicators effectively captured realistic trends in biodiversity caused by a policy change, that is, whether available indicators are useful in detecting policy relevant trends.

This study provides an opportunity to assess the feasibility of scenario modeling and its application to protected area management. The scenarios explored in this study were as follows:

Scenario 1: "Business as usual" assumed current population trends in protected areas for each species in each region were applied to populations inside existing protected areas (drawn from the WDPA) (WDPA, 2010). Trends outside protected areas were assumed to be 25% worse than those inside protected areas (Nicholson et al., 2012; Costelloe et al., 2015).

Scenario 2: Expand terrestrial protected area coverage to 10% of each country. The conservation planning software Marxan (Possingham et al., 2000) was used to select 25,625 km^2 grid cells to add to the current protected area network. These were based on habitat suitability models for the target species (Rondinini et al., 2005). Country-level targets of 10% of each country were used (one of the original targets set for global protected area coverage by the Convention on Biological Diversity (CBD, 2010b)). It was assumed that a suitable habitat added to the protected areas was occupied (thus potentially overestimating the positive impacts of expansion due to commission errors) and had the same population density as that in the current protected areas. The population trends for each species in each region applied to populations inside and outside protected areas were the same as those used in "business as usual" scenario (Scenario 1).

Scenario 3: Expand terrestrial protected area coverage to 17% of each country, as per Scenario 2 except to a country-level target of 17% (CBD, 2010a).

Scenario 4: Improved management effectiveness in protected areas. No consistent data exist on the impact of effective management within protected areas across Africa; however, protected areas in southern Africa are considered to be the most effectively managed (Craigie et al., 2010). Therefore, in order to simulate effective management, it was assumed that populations in effectively managed protected areas underwent the same observed annual rate of increase as the average across all species in southern African protected areas (+1.8%) (Craigie et al., 2010), except for those that already had a more positive annual trend, the rate of which was assumed would stay constant. A sensitivity analysis showed that the value of the assumed trend in effectively managed protected areas had no effect on the relative impacts of the policies (Costelloe et al., 2015). Population trends outside protected areas were the same as those used for in "business as usual" scenario (Scenario 1).

Scenario 5: Expand protected area coverage to 10% of each country and increase management effectiveness of protected areas. The areas added to the current protected area network were the same as those described earlier in Scenario 2, with the same corresponding redistribution of populations between protected areas and nonprotected areas. The population trends inside protected areas were the same as those used in the effectiveness scenario (Scenario 3), based on current trends in southern Africa. Trends outside protected areas were the same as "business as usual" scenario (Scenario 1).

Scenario 6: Expand protected area coverage to 17% of each country and increase management effectiveness of protected areas—as per Scenario 5 except to a country-level target of 17%.

Two indicators were used to track changes in wildlife in protected areas: the IUCN Red List Index (Butchart et al., 2004) and the Living Planet Index (Loh et al., 2005). These were chosen by the study authors as both are well developed and data rich and measure subtly different aspects of biodiversity, both important to assessing progress toward the CBD targets (Walpol et al., 2009) and to evaluating protected area management (Nicholson et al., 2012). The analyses comprised a series of three steps:

1. Establish estimates of population size, distribution, and current rates of decline for each species.
2. Grow (or reduce) the mammal populations under each of the policy scenarios.
3. Measure resultant changes in species abundance under the scenarios using the two indicators.

The case study comprised 53 mammal species in 41 countries divided among four regions: east, southern, west, and central Africa. Country-level estimates of population size for each species inside and outside of protected areas were collated or estimated from the literature (predominantly the IUCN Species Survival Commission publications). For each species in each region, interannual trends in population sizes within protected areas were estimated from time series as described by Craigie et al. (2010) and were modeled as set out above, using a GAM or simple chain method. Where multiple estimates of trends were available within a region, such as several protected areas in one country or across several countries, the geometric mean of population trends was used to produce a regional trend per species in protected areas. Currently, no comprehensive trend data exist for the study species outside protected areas, although empirical data suggest lower densities of mammal species outside protected areas than inside Tanzania (Caro, 1999; Struhsaker et al., 2005; Setsaas et al., 2007).

In the absence of appropriate trend data, it was assumed that population trends for all species would be 25% worse outside protected areas than inside protected areas. Therefore, positive trends were decelerated by 25% and negative trends accelerated by 25%. Sensitivity analyses showed the qualitative results to be robust to the assumed difference in trend between populations inside and outside protected areas (Costelloe et al., 2015). The impact of each policy scenario on trends in abundance of the study species was modeled over a 30-year period, assuming constant annual trends and immediate implementation.

Values for the two indicators were calculated under each scenario from the population projections detailed above. The Red List Index was calculated for each scenario at decadal intervals (Butchart et al., 2004, 2007). The Living Planet Index was calculated annually over the time period (Collen et al., 2009). For more details on methods and sensitivity analyses, see Costelloe et al. (2015).

Both of the indicators (abundance and extinction risk) proved to be capable of discerning between different future protected area policy scenarios. Overwhelmingly, both indicators showed that without effective management, expanding protected areas provided negligible benefit over the "business as usual" scenario (Figure 6.3a and b) (Nicholson et al., 2012; Costelloe et al., 2015). There was little difference between expanding national protected area coverage to 10 or 17% and expansion in conjunction with increased management effectiveness provided little benefit above simply increasing effectiveness of the existing network. The model outputs demonstrated that differentiation between rankings of each policy scenario was very similar for the two indicators. There were differences in the magnitude of the differences they displayed, which reflect the manner in which the indicators deal with underlying abundance data. For example, the Red List Index was stable or marginally increased under the improved management effectiveness scenarios and declined over the first decade before starting slow recovery under the scenarios without improved management effectiveness (Figure 6.3a). By contrast, the Living Planet Index declined at an attenuating rate, with more severe declines apparent for the scenarios without improved management effectiveness (Figure 6.3b).

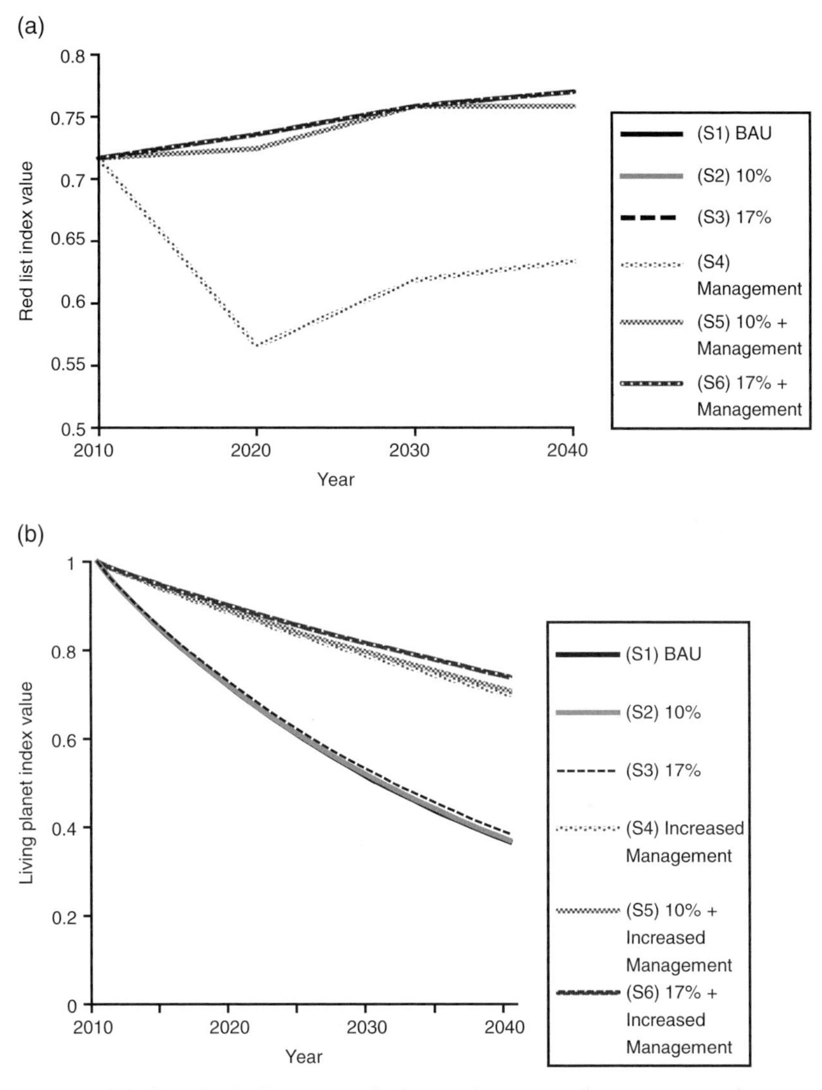

Figure 6.3 **Biodiversity indicators track changes in protected area status under a range of different management scenarios. (a) Red List Index. Source: From Nicholson et al. (2012). (b) Living Planet Index. Source: From Costelloe et al. (2015). Reproduced under the terms of the Creative Commons Attribution License CC BY 3.0**

While both indicators tested appear reliable in reflecting general trends produced by hypothesized policy interventions, part of the explanation for the difference in indicator trajectory is revealed by single species examples. Costelloe et al. (2015) use the tsessebe (*Damaliscus lunatus*) population trend from the "business as usual" scenario to exemplify the impact of changing distribution of abundance across the range of the species, which is typical of species populations with this type of decline dynamic. For this species, at the start of the scenario (2010), there was a large east African population subject to a particularly strong annual decline (−13.4% per year) that drove the overall population down sufficiently (−87% in the first decade) for reclassification from least concern to critically endangered under criterion A, in the Red List Index. For the Living Planet Index, this decline is reflected as a rapid declining trend. Over time, the declining east African population comprised a smaller proportional share of the continent-wide population, with the previously smaller southern African population becoming proportionately larger (a positive trend of +1.3% per year). By the final decade of the model, the southern African population comprised the majority of the total pan-African population of the species (albeit heavily depleted), resulting in reclassification to least concern in latter decades for the Red List Index, and an attenuating decline for the Living Planet Index. The resulting indices look rather different; the Red List Index based on extinction risk rebounds to its 2010 value in 2040, accurately reflecting extinction risk as estimated with the Red List Index criteria (IUCN, 2001) but losing the existence of historically higher abundance. The Living Planet Index portrays a far lower abundance in 2040 and reflects the change from an historic high. The main insight this study suggests is that we must choose our indicators with caution and make sure that they reflect a metric of biodiversity that we value.

Discussion

Existing protected area monitoring

Though broad in geographic coverage (133 countries are represented), the dataset of abundance time series trends in protected areas presented here clearly shows bias in both taxonomic and geographic representation. Foremost, there are large gaps in monitoring data from protected areas in South America and Southeast Asia which contrasts poorly with current knowledge about hotspots of species diversity (Jenkins et al., 2013). The data suggest that a small proportion of the world's wildlife populations in protected areas are being monitored. Further, we evaluated only vertebrate species coverage, and clearly, protected areas are refuge to more than just vertebrates (D'Amen et al., 2013). This limits the ability to make robust conclusions about the trends and long-term persistence of species in protected areas. While search efforts for abundance time series data for the Living Planet Index database used in this analysis have not explicitly targeted protected areas other than those collated for Craigie et al. (2010), we estimate that 18–20 researcher years of effort have been put into collating data for this database (excluding the efforts made by researchers and park authorities for the original monitoring, which would significantly increase this estimate).

Regional case studies, like the African study presented, show however that with concerted effort, it is possible to get a broad picture of trends in protected areas, if sufficient existing monitoring schemes can be brought together. These results show troubling signs for mammals in African protected areas. There has evidently been a continent-wide halving in the abundance of large mammals in African reserves since the 1970s (Craigie et al., 2010) with overarching causes thought to be hunting and habitat conversion. There are of course important caveats. First, this type of analysis lacks data on the population trends of wildlife in the areas surrounding the protected areas from which trends are measured. The effectiveness of protective measures compared with what might have happened in their absence is unknowable, unless monitoring is carried out beyond park borders. Yet these data, and data on the effectiveness of different management types, are essential for evidence-based protected area management.

More than 150,000 protected areas have been designated around the world, with a shared goal of protecting species and ecosystems. Yet only a fraction of these protected areas appear to have wildlife monitoring systems in place (Coad et al., 2013). That they can continue to achieve their goal as human impact escalates in severity and extent remains a question of broad concern and a key focus of conservation research.

Establishing and strengthening monitoring

Indicators of changing biodiversity are only as good as the data underlying them, so careful thought needs to go into enhancing monitoring. While the only broadscale case study available to our knowledge (for African protected areas) shows worryingly negative trends, our ability to take into account counterfactual change in biodiversity trends beyond park boundaries is almost nonexistent (Ferraro, 2009). There are a number of ways in which protected area monitoring could be improved.

Firstly, monitoring must be expanded in geographic terms. Tropical areas are noted to have the have fastest instantaneous rate of change in biodiversity and yet are consistently shown to be the lowest in data coverage in a number of metrics of biodiversity, not least abundance (Collen et al., 2008, 2009; Tobias et al., 2013). Secondly, the counterfactual trend in species beyond park borders should be measured. Of course, one might argue that enough is known (we know that biodiversity is in decline, so we should just act) or that sometimes due to the reality of limited resources for conservation, the correct decision may be not to monitor at all (Jones, 2013). Nevertheless, protected areas lie at the center of biodiversity protection plans, and large amounts of resources are dedicated to their proper functioning, so we ought to be able to measure the extent to which they are working (Fuller et al., 2010).

Continent-wide programs such as Monitoring the Illegal Killing of Elephants (Hedges & Lawson, 2006) have been set up to provide information needed for countries with elephant populations to make appropriate management and enforcement decisions, as well as build long-term institutional capacity. Such programs might provide a model from which a broader species coverage comparing trends in protected to nonprotected areas can be achieved. Coupled species and population level data will be a requirement for metrics of management effectiveness (Geldmann et al., 2013). Identification of the most appropriate measures for effectiveness remains elusive, but candidate metrics are apparent. Whether trends of species in protected areas are distinguishable from those in the surrounding lands is of fundamental importance to science and to protected area protection policy. Evaluating the efficacy of protected areas over time should be a focus for the future, as available data for large vertebrates suggests that populations outside protected areas are significantly lower than those inside. Thirdly, species coverage is limited. The database that we evaluated contains no invertebrates or plants, and if society is interested in what protected areas provide in terms of ecosystem processes and services, as well as the financial value that comes from activities such as tourism, invertebrates are no doubt of greater importance than current meager level of monitoring suggests.

Current indicators and realistic policy options

Of the global indicators of biodiversity change that are available, the two tested that were presented here (the Living Planet Index and the Red List Index) are probably the best developed. Both appear to be able to qualitatively reflect differences in policy under the scenarios that were applied, as well as illustrate past trends. The finding that both reflect overwhelmingly that without effective management, expanding protected areas provided negligible benefit over the "business as usual" scenario is a powerful finding for protected area policy. It raises further issues of global biodiversity targets and shows the potential for mismatch of politically derived targets (the CBD target is for a threshold coverage of 17% the global land surface to be protected by 2020) and potentially more effective policy mechanisms (in this scenario, enhancing management impact). As Costelloe et al. (2015) point out, there are many assumptions, simplifications, and caveats to this type of scenario modeling, which must be rigorously tested and refined. However, without sufficient information on biodiversity within and around protected areas, they are perhaps one of our best and most immediate forms of recourse for making more informed policy decisions.

Acknowledgments

We thank Lucas Joppa, Megan Barnes, Stephen Woodley, and the IUCN Protected Area Task Force for discussion; Charlotte Outhwaite and Stefanie Deinet for providing summary statistics on the Living Planet Index database; Brendan Costelloe and Kate Sullivan for analyses; and the Zoological Society of London for hosting the conference that led to this work. BC was partially supported by the Rufford Foundation, LM by WWF International, SW by the EU BIOPAMA project, EJMG by a Royal Society Wolfson Research Merit Award, EN by a Marie Curie Fellowship, and IDC through NERC studentship NER/S/A/2006/14094 with CASE support from UNEP-WCMC.

References

Balmford, A., Green, R. E., & Jenkins, M., 2003. Measuring the changing state of nature. *Trends in Ecology & Evolution*, Volume 18, p. 27–31.

Beale, C. M., Baker, N. E., Brewer, M. J., & Lennon, J. J., 2013. Protected area networks and savannah bird biodiversity in the face of climate change and land degradation. *Ecology Letters*, Volume 16, p. 1061–1068.

BirdLife International, 2011. BirdLife International Taxonomic Checklist Version 4.0. [Online] Available at: www.birdlife.org/datazone/info/taxonomy (Accessed July 11, 2015).

Buckland, S. T., Magurran, A. E., Green, R. E., & Fewster, R. M., 2005. Monitoring change in biodiversity through composite indices. *Philosophical Transactions of the Royal Society of London Series B*, Volume 360, p. 243–254.

Butchart, S. H. M. et al., 2004. Measuring global trends in the status of biodiversity: Red list indices for birds. *PLoS Biology*, Volume 2(12), p. 2294–2304.

Butchart, S. H. M. et al., 2007. Improvements to the red list index. *PLoS ONE*, Volume 2(1), p. e140.

Butchart, S. H. M. et al., 2012. Protecting important sites for biodiversity contributes to meeting global conservation targets. *PLoS ONE*, Volume 7(3), p. e32529.

Caro, T. M., 1999. Densities of mammals in partially protected areas: The Katavi Ecosystem of western Tanzania. *Journal of Applied Ecology*, Volume 36, p. 205–217.

CBD, 2010a. *Decisions adopted by the CoP to the CBD at its 10th meeting (UNEP/CBD/COP/DEC/X/2)*, Montreal: Secretariat of the CBD.

CBD, 2010b. *Global biodiversity outlook 3*, Montreal: Secretariat of the CBD.

Ceballos, G., & Ehrlich, P. R., 2002. Mammal population losses and the extinction crisis. *Science*, Volume 296, p. 904–907.

Chape, S., Harrison, J., Spalding, M., & Lysenko, I., 2005. Measuring the extent and effectiveness of protected areas as an indicator for meeting global biodiversity targets. *Philosophical Transactions of the Royal Society of London Series B*, Volume 360(1454), p. 443–455.

Chessman, B. C., 2013. Do protected areas benefit freshwater species? A broad scale assessment for fish in Australia's Murray-Darling Basin. *Journal of Applied Ecology*, Volume 50, p. 969–976.

Coad, L. et al., 2013. Progression towards the CBD protected area management effectiveness targets. *PARKS*, Volume 19(1), p. 13–24.

Collen, B., Ram, M., Zamin, T., & McRae, L., 2008. The tropical biodiversity data gap: Addressing disparity in global monitoring. *Tropical Conservation Science*, Volume 1(2), p. 75–88.

Collen, B. et al., 2009. Monitoring change in vertebrate abundance: The living planet index. *Conservation Biology*, Volume 23, p. 317–327.

Collen, B. et al., 2011. Predicting how populations decline to extinction. *Philosophical Transactions of the Royal Society of London Series B*, Volume 366, p. 2577–2586.

Collen, B. et al., 2013. Tracking changes in abundance: The living planet index. In: B. Collen, N. Pettorelli, J. E. M. Baillie, & S. M. Durant, eds. *Biodiversity monitoring and conservation: Bridging the gap between global commitment and local action*. Chichester: John Wiley & Sons, Ltd, p. 71–94.

Costelloe, B. et al., 2015. Global biodiversity indicators reflect the modelled impacts of protected area policy change. *Conservation Letters*. doi: 10.1111/conl.12163.

Craigie, I. D. et al., 2010. Large mammal population declines in Africa's protected areas. *Biological Conservation*, Volume 143, p. 2221–2228.

D'Amen, M. et al., 2013. Protected areas and insect conservation: Questioning the effectiveness of Natura 2000 network for saproxylic beetles in Italy. *Animal Conservation*, Volume 16(4), p. 370–378.

Ehemann, C., 2007. Evaluating and adjusting for chain drift in national economic accounts. *Journal of Economics and Business*, Volume 59, p. 256–273.

Ferraro, P. J., 2009. Counterfactual thinking and impact evaluation in environmental policy. *New Directions for Evaluation*, Volume 2009, p. 75–84.

Ferraro, P. J., & Pattanayak, S. K., 2006. Money for nothing? A call for empirical evaluation of biodiversity conservation investments. *PLoS Biology*, Volume 4(4), p. e105.

Fewster, R. M. et al., 2000. Analysis of population trends for farmland birds using generalize additive models. *Ecology*, Volume 8, p. 1970–1984.

Frisch, R., 1936. Annual survey of economic theory: The problem of index numbers. *Econometrica*, Volume 4, p. 1–39.

Fuller, R. A. et al., 2010. Replacing underperforming protected areas achieves better conservation outcomes. *Nature*, Volume 466, p. 365–367.

Geldmann, J. et al., 2013. Effectiveness of terrestrial protected areas in reducing habitat loss and population declines. *Biological Conservation*, Volume 161, p. 230–238.

Gregory, R. D. et al., 2005. Developing indicators for European birds. *Philosophical Transactions of the Royal Society of London Series B*, Volume 360, p. 269–288.

Hedges, S., & Lawson, D., 2006. *Dung survey standards for the MIKE programme*, Nairobi: CITES MIKE.

IUCN, 2001. *IUCN red list categories and criteria: Version 3.1*, Gland & Cambridge: IUCN SSC.

Jenkins, C. N., Pimm, S. L., & Joppa, L. N., 2013. Global patterns of terrestrial vertebrate diversity and conservation. *Proceedings of the National Academy of Sciences of the United States of America*, Volume 110(28), p. E2602–E2610.

Jones, J. P. G., 2013. Monitoring in the real world. In: B. Collen, N. Pettorelli, J. E. M. Baillie, & S. M. Durant, eds. *Biodiversity monitoring and conservation: Bridging the gap between global commitment and local action*. Chichester: John Wiley & Sons, Ltd, p. 335–347.

Loh, J. et al., 2005. The living planet index: Using species population time series to track trends in biodiversity. *Philosophical Transactions of the Royal Society of London Series B*, Volume 360(1454), p. 289–295.

Mace, G. M. et al., 2008. Quantification of extinction risk: IUCN's system for classifying threatened species. *Conservation Biology*, Volume 22, p. 1424–1442.

McKee, J., Chambers, E., & Guseman, J., 2013. Human population density and growth validated as extinction threats to mammal and bird species. *Human Ecology*, Volume 41(5), p. 773–778.

McRae, L. et al., 2007. *A living planet index for Canada*, Gland: WWF.

McRae, L. et al., 2012. The arctic species trend index: Using vertebrate population trends to monitor the health of a rapidly changing ecosystem. *Biodiversity*, Volume 13, p. 144–156.

Living Planet Index, 2012. *The living planet index data portal*. London: ZSL & WWF.

NERC Centre for Population Biology, 2010. Dynamics Database Version 2. [Online] Available at: http://www.sw.ic.ac.uk/cpb/gpdd.html (Accessed July 11, 2015).

Newman, W. D., 2008. Isolation of African protected areas. *Frontiers in Ecology and the Environment*, Volume 6, p. 321–328.

Nicholson, E. et al., 2012. Making robust policy decisions using global biodiversity indicators. *PLoS ONE*, Volume 7(7), p. e41128.

Pereira, H. M. & Cooper, H. D., 2006. Towards the global monitoring of biodiversity change. *Trends in Ecology and Evolution*, Volume 21, p. 123–129.

Pomeroy, D. & Tushabe, H., 2010. *The state of Uganda's biodiversity 2010*, Kampala: University Institute of Environment and Natural Resources.

Possingham, H., Ball, I., & Andelman, S., 2000. Mathematical methods for indentifying representative reserve networks. In: S. Ferson & M. Burgman, eds. *Quantitative methods for conservation biology*. New York: Springer, p. 291–306.

Rondinini, C., Stuart, S., & Boitani, L., 2005. Habitat suitability models reveal shortfall in conservation planning for African vertebrates. *Conservation Biology*, Volume 19, p. 1488–1497.

Scholte, P., 2011. Towards understanding large mammal population declines in Africa's protected areas: A west-central African perspective. *Tropical Conservation Science*, Volume 4(1), p. 1–11.

Setsaas, T. H. et al., 2007. How does human exploitation affect impala populations in protected and partially protected areas? A case study from the Serengeti Ecosystem. *Biological Conservation*, Volume 136, p. 563–570.

Struhsaker, T., Struhasaker, P., & Siex, K., 2005. Conserving Africa's rainforests: Problems in protected areas and possible solutions. *Biological Conservation*, Volume 123, p. 45–54.

Tobias, J. A., Sekercioglu, C. H., & Vargas, F. H., 2013. Bird conservation in tropical ecosystems. In: D. W. MacDonald & K. Willis, eds. *Key topics in conservation biology*. Oxford: Blackwell, p. 258–276.

Walpol, M. et al., 2009. Tracking progress towards to the 2010 biodiversity target and beyond. *Science*, Volume 325, p. 1503–1504.

WDPA, 2010. *The world database on protected areas (WDPA)*, Gland & Cambridge: IUCN, UNEP-WCMC.

Western, D., Russell, S., & Cuthill, I., 2009. The status of wildlife in protected areas compared to non protected areas in Kenya. *PLoS ONE*, Volume 4(7), p. e6140.

7

Effectiveness of Protected Areas in Conserving Large Carnivores in Europe

L. Santini, L. Boitani, L. Maiorano and C. Rondinini

Department of Biology and Biotechnologies,
Università di Roma, Rome, Italy

Introduction

Europe is a densely populated continent and has been profoundly shaped by human presence since the last glaciation. Large wilderness areas are absent from Europe and land cover is mostly dominated by human activities (Linnell et al., 2005). As a consequence, the conservation strategies of most European countries have been based on a combination of small protected areas (PAs) and the integration of biodiversity and human activities in both natural and human-modified habitats (Redford et al., 2003). These strategies are implemented through national and international legislation that establishes PAs and set restrictions to anthropic activities outside those areas (Rondinini & Boitani, 2007).

In addition to national protected areas (NPAs), Europe has implemented a network of conservation areas called Natura 2000 (Evans, 2012). Natura 2000 is arguably the most important conservation tool in Europe and is regulated by two directives: the 1979 Bird Directive and the 1992 Habitat Directive. The Bird Directive aimed to the identification of sites for the conservation of bird species (Special Protection Areas (SPAs)), whereas the Habitat Directive aimed to conserve other species and habitats through the identification of sites of community importance (SCIs) that, if designated, would have become Special Areas of Conservation (SACs). SPAs and SACs together compose the Natura 2000 network, which aims to conserve European biodiversity through a system of connected conservation sites that adds to the existing NPAs, complementing them in case of their mismanagement or removal. Natura 2000 is the largest assemblage of PAs in

Protected Areas: Are They Safeguarding Biodiversity?, First Edition.
Edited by Lucas N. Joppa, Jonathan E. M. Baillie and John G. Robinson.

the world and is composed by more than 26,000 sites covering about 17.5% of the EU land territory (Evans, 2012).

The Natura 2000 network was implemented independently from national strategies so there is a substantial overlap with NPAs. In addition, sites identification in the Natura 2000 network has been completely devolved to single countries and no general, EU-wide, spatially explicit conservation planning strategy was considered (Rondinini & Pressey, 2007). All areas proposed by individual countries were then discussed and approved by national experts and the European Commission during 'biogeographical seminars', with no clear criteria for approval. As a consequence, each country determined size and distribution of sites with different strategies (i.e. some countries established a few large sites while others established many small sites; Evans, 2012), and the contribution of the overall network to specific conservation objectives needs to be assessed *a posteriori*. Natura 2000 sites may not be considered fully PAs, however for simplicity, here after we will refer to PAs for both NPAs and Natura 2000 sites.

Past evaluations of the European PAs (Maiorano et al., 2006, 2007, 2015; Araùjo et al., 2007; Jantke et al., 2011) used more or less standard gap analyses to assess the amount of species distribution covered by PAs, with targets set on the basis of fixed or proportional representation (e.g. proportion of the species' geographic range included in PAs). To achieve long-term species conservation, it is indeed necessary to conserve viable populations, which are difficult to estimate. The extent of occurrence is not a good proxy of the area occupied by species, as it depends on the species' habitat specificity (Rondinini et al., 2011). In addition, even the amount of occupied area alone does not inform about species persistence (i.e. number and size of viable populations), unless its spatial arrangement and species characteristics are accounted for. This may explain why only a limited number of studies so far have tried to incorporate considerations of population viability in the evaluation of PAs (Allen et al., 2001; Pinto & Grelle, 2009; Jantke et al., 2011; Santini et al., 2014).

This study aims to evaluate the ability of the PA system to conserve viable populations of the three largest carnivores in Europe: the lynx (*Lynx lynx*), the wolf (*Canis lupus*) and the bear (*Ursus arctos*). Direct persecution combined with deforestation and decline in ungulate prey led European large carnivores to local extinction in many areas (Linnell et al., 2001). Legislation, reforestation, prey population increase and successful supplementations and reintroductions determined a change in this negative trend since the 1960s (Boitani, 2000; Breitenmoser et al., 2000; Swenson et al., 2000; Linnell et al., 2005; Chapron et al., 2014). However, because of the difficulties related to well-known conflicts with human activities (Treves & Karanth, 2003; Rondinini & Boitani, 2007) and their large area requirements, large carnivores in Europe are still a conservation challenge (European Commission, 2007; Linnell et al., 2008). Only recently, Linnell et al. (2008) underlined the importance of a population approach for large carnivore conservation and identified distinct (sub-)populations of large carnivores in Europe.

The role of the PA network for large carnivore conservation in Europe has not yet been evaluated. In fact, except for a few countries (e.g. Spain and Sweden), Natura 2000 is mostly composed of very small areas (Linnell et al., 2005; Gaston et al., 2008; Evans, 2012; Maiorano et al., 2015), which are individually unlikely to be effective for the long-time persistence of large carnivores (Linnell et al., 2008; Santini et al., 2014).

We consider both NPAs and SACs and the contribution of the two separately. We account for the spatial arrangement of the suitable areas within species' geographic ranges, their average population density and their dispersal abilities in order to infer their likely spatial structuring in distinct populations (Santini et al., 2014). We consider the minimum amount of area required per site and their interconnection according to the species dispersal abilities. For each species, we develop two spatial models that incorporate all these variables according to two dispersal scenarios: an optimistic scenario where species are assumed to generally disperse long distances and to be able to settle in small areas and a more conservative scenario where species are assumed to disperse shorter distances and to need larger areas to settle in.

We then compare the total coverage provided by PAs to species' suitable habitat to the coverage provided to potential viable populations. We also assess the contribution of each individual European country to large carnivore conservation and their ability to independently conserve the species at a national level through their PAs and SACs.

Methods

Population distribution models

The models were based on the global habitat suitability models for all terrestrial mammals from Rondinini et al. (2011). These expert-based models assign three suitability levels: high (i.e. primary habitat of a species), medium (i.e. secondary habitat where a species can be found but not persist in the absence of primary habitat) and low suitability (i.e. where species are expected to be seldom or never found) (Rondinini et al., 2011). In our models, medium suitability areas were considered as highly suitable when they laid within a species home range diameter from high suitability areas, in accordance to the definition of medium suitability. The remnant medium suitability areas were considered unsuitable.

We defined minimum patch size (MPS) as the minimum area needed to sustain a few individuals in order to allow reproduction and a second generation of dispersers. We considered two different MPS according to the two dispersal scenarios ('optimistic' and 'conservative' see succeeding texts), and all suitable patches smaller than the MPS were removed. All suitable patches smaller than the MPS were removed. These patches may be important for dispersal dynamics by providing important 'stepping stones' but were irrelevant to our models because we

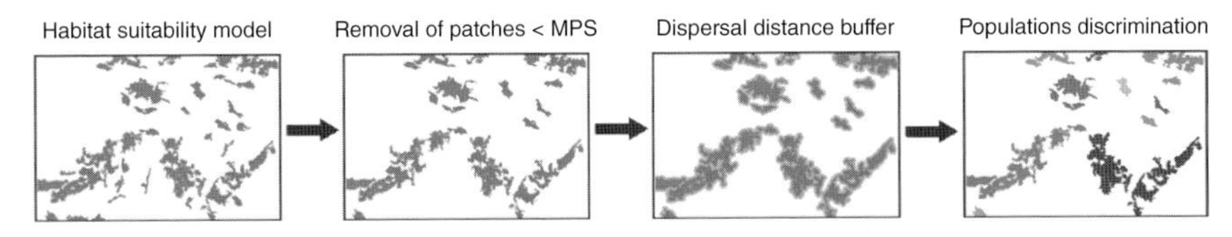

Figure 7.1 **Main steps followed in the development of population distribution models**

assumed that dispersal was possible in a buffer equal to the Euclidean dispersal distance of females, irrespective of stepping stones. All patches larger than the MPS and within a female dispersal distance were considered part of the same cluster (i.e. group of potentially connected patches; Swenson et al., 2000). In fact, dispersal distance is generally shorter in females among mammals (Greenwood, 1980), and for the purpose of this study, we considered two populations as distinct if unable to recolonize them reciprocally (Figure 7.1). A similar approach to the identification of distinct populations was used by Santini et al. (2014).

PA network models

The Natura 2000 vector map was downloaded from http://eunis.eea.europa.eu/. Natura 2000 is composed of SPAs (Birds Directive 1979) and SACs (Habitat Directive 1992); however, several sites of the network are considered both SPAs and SACs. All exclusively SPA sites (type codes A, D, F, H, J) were removed from the network because they aimed at protecting bird species.

The NPA vector map was downloaded from http://www.protectedplanet.net/. We selected only PAs in IUCN classes from I to IV (IUCN, 1994). For all the PAs represented as points (i.e. where PA shape is not specified), we generated a buffer equal to the radius of the PA reported area.

The vector maps were then projected to Mollweide equal-area projection and rasterized at 300 m resolution. Cells were considered protected if more than 50% were inside PAs.

The same modelling framework applied to habitat suitability models (Figure 7.1) was applied to NPA and SAC sites, both alone and together. All the sites smaller than the species MPS were removed by the network. All remnant sites within a species dispersal distance were considered part of a cluster (i.e. group of potentially connected protected sites). Protected sites were considered as occupied by species if suitable and within their extent of occurrence (Maiorano et al., 2007).

The analysis window for NPAs and NPAs+SACs included also PAs in non-EU countries west to the eastern limit of EU in Romania (i.e. Norway, Switzerland, Bosnia and Herzegovina, Croatia, Montenegro, Albania, Moldavia, Belarus and part of Russia and Ukraine). This was necessary to avoid the error of considering some EU PAs not connected because non-EU PAs between them were excluded.

Dispersal scenarios

In order to encompass the uncertainty related to MPS and dispersal parameters, two scenarios were considered for both species' population and species' PA network models.

In the 'optimistic' scenario, MPS was considered equal to two home range areas for the lynx and the bear and one pack home range area for the wolf. Patches were considered connected if within a maximum dispersal distance.

In the 'conservative' scenario, the MPS corresponded to 10 home range areas for the lynx and the bear and two pack home range areas for the wolf. Patches were connected if within 1/5 of the maximum dispersal distance, approximately corresponding to median dispersal distance (Table 7.1; Santini et al., 2013).

Analyses

We first measured the percentage of suitable area covered by PAs for each species at the European level (baseline coverage). We then quantified the percentage of protected suitable area accounting for a minimum patch size according to the two dispersal scenarios (MPS coverage). Finally, we calculated the percentage of protected suitable area reaching the population conservation target (effective coverage). The population conservation target was set at 500 individuals for the wolf (Liberg, 2006) and the lynx (U. Breitenmoser, personal communication) and at 250 for the bear (Wielgus, 2002). These may be considered as demographic MVPs in the absence of threatening processes, that is, minimum population sizes robust to demographic stochasticity; if threats such as poaching are present, the MVP size may have to be much higher. The population conservation target was converted in area target dividing it by the species median population density (i.e. 0.02 individuals/km^2 for all species (Jones et al., 2009): 25,000 km^2 for lynx and wolves and 12,500 km^2 for bears). We determined the achievement of the targets at

Table 7.1 **Species life history data used to develop the population and protected area network models**

	Home range (km^2)	Density (n/km)	Maximum recorded dispersal distance for females (km)
Lynx lynx	136.35	0.02	82
Canis lupus	250	0.02	1092
Ursus arctos	310.81	0.02	90

Source: Data from Jones et al. (2009), Støen et al. (2006), Wabakken et al. (2007), and Zimmermann et al. (2005).

the intersection between the populations (as identified in the population models) and clusters of connected protected sites, including neighbouring NPAs in non-EU countries (see 'PA network models' paragraph). We also tested whether each individual site (NPAs and SACs) was able to support viable populations in isolation. Finally, for each species, we calculated the amount of protected suitable area in each EU country according to the two scenarios, its potential supported number of individuals and the relative contribution to the species overall protection in EU. We assessed which countries are able to sufficiently protect viable populations of the three species within their boundaries.

Analyses were conducted on Grass v. 6.4 (GRASS Development Team, 2012) and R v. 2.15.2 (R Core Team, 2012).

Results

The suitable habitat of the three species in Europe is covered between 5 and 10% by NPAs and between 15 and 20% by SACs. The percentage is only marginally higher when NPAs and SACs are considered together because the two overlap (Figure 7.2).

When all the protected sites smaller than the MPS are excluded from the analysis, the percentage of suitable area in PAs drops considerably, especially for the Natura 2000 sites (Figure 7.2).

According to the two dispersal scenarios, the lynx distribution in Europe is structured in six to seven populations (i.e. clusters of connected suitable patches), one to two of which results sufficiently protected. The wolf is clustered in one population in the optimistic dispersal scenario and in four populations in the conservative dispersal scenario. In the former scenario, the population is sufficiently protected by one cluster of protected sites, and in the latter, it is sufficiently protected by two clusters in the same population. The bear is clustered in four and nine populations in the optimistic and conservative scenario, 1 and 2 of which are sufficiently protected, respectively.

Except for one SAC in Sweden, which is large enough to support a viable population of bear alone (1.7% of its protected habitat), no single protected site is sufficient to sustain a viable population of any of the three large carnivores.

Considering NPAs and SACs together, lynx habitat is mostly protected in Sweden (32.4%), Finland (27.8%) and Romania (13.2%). The countries which provide more protection for the wolf are Spain (16.4%), Romania (13.9%), Bulgaria (11.7%) and Italy (10.6%). Bear is mostly protected in Sweden (28.4%), Finland (26.9%), Romania (11.7%) and Bulgaria (7.5%). Considering the spatial structuring of the populations and the suitable habitat within protected sites, the only countries able to support viable populations of large carnivores are Sweden and Finland for the lynx and the bear and Romania for the bear in the optimistic

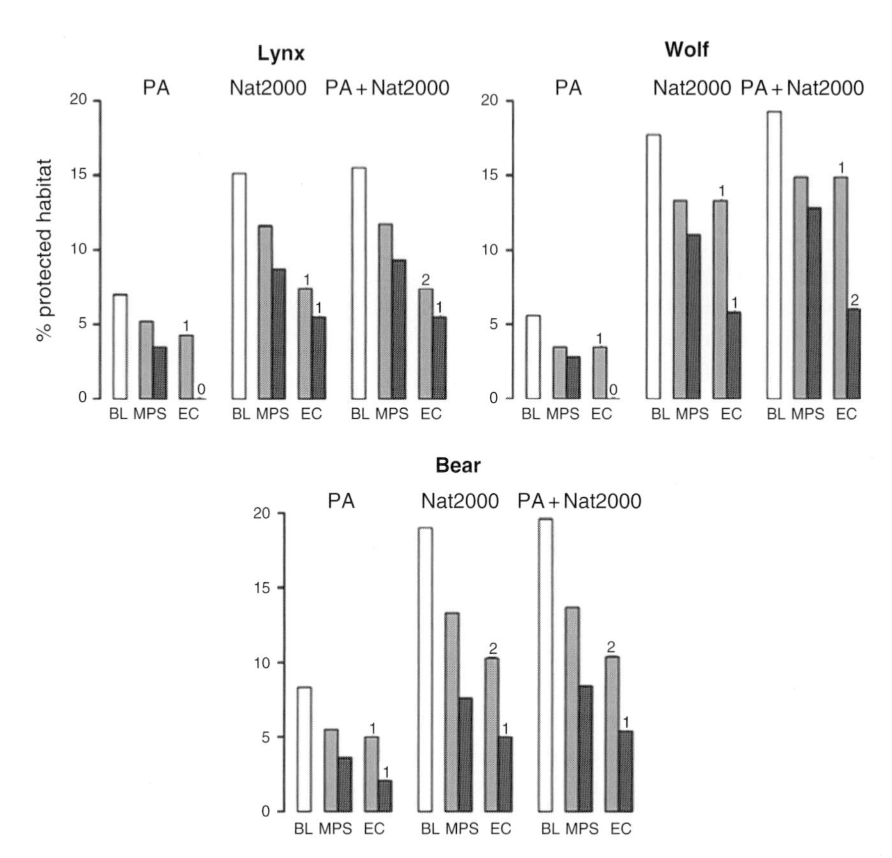

Figure 7.2 **Coverage percentage of protected areas for three species in Europe. BL, baseline coverage (protected habitat/overall habitat); EC, effective coverage (protected habitat reaching the target at the cluster level/overall habitat); MPS, MPS coverage (protected habitat with a minimum patch size/overall habitat); NPAs, national protected areas; SACs, Special Areas of Conservation in Natura 2000 network. Values above the EC bars indicate the number of protected site clusters able to support viable populations. Source: Data from Natura 2000 (2013), WDPA (2010) and IUCN (1994)**

dispersal scenario. However, if we assumed that all PAs within a country are occupied and connected (i.e. not considering MPS and dispersal distance), also Spain would support a viable population of the wolf (Figure 7.3).

Discussion

In this study, we assessed the ability of the NPAs and Natura 2000 network to provide sufficient coverage for the three largest European carnivores. Their conservation is in fact of great concern both because of the intrinsic difficulty in managing

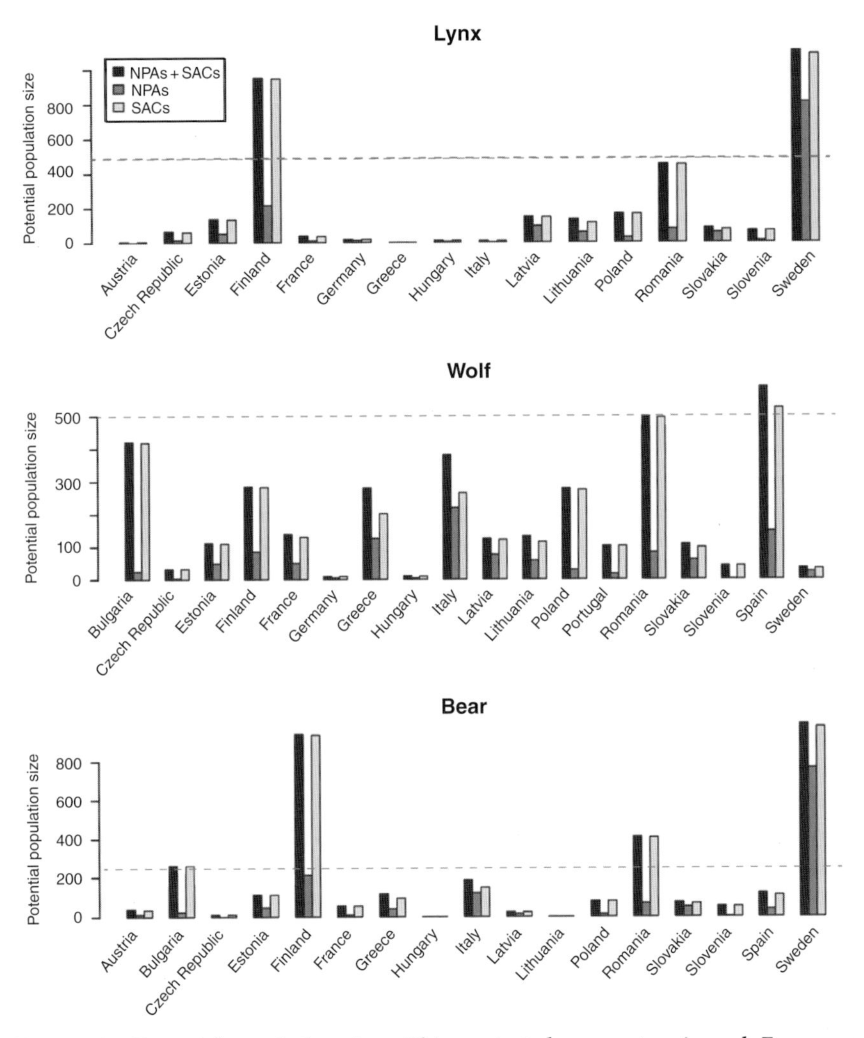

Figure 7.3 **Potential population size within protected area system in each European country (assuming their occupation and connection within each country and considering an average density of 0.02 n/km²). Dashed lines represent the target for the species. Source: Data from Natura 2000 (2013), WDPA (2010) and IUCN (1994)**

their large spatial needs and their conflicts with human activities (Treves & Karanth, 2003; European Commission, 2007; Rondinini & Boitani, 2007; Linnell et al., 2008; Chapron et al., 2014).

Our study was conducted under three main assumptions. The first is that PAs are well managed and effective in conserving the study species, which may not be realistic. In many cases, conservation action in Natura 2000 sites is limited to just publishing regulations limiting human activities – there is no active management or enforcement. The second assumption is that our models are reliable representations of the area occupied by species and of their structure in distinct populations. In fact, even if we added a new dimension that helps refining the habitat suitability models, our models still overestimate occupation and connectivity. This results because the habitat suitability models that we used as a baseline only account for suitable habitat (Rondinini et al., 2011) and ignore factors such as human dimension or prey availability, which are crucial for large carnivore presence (Woodroffe, 2000; Carbone & Gittleman, 2002). The third assumption is that the matrix is homogeneously permeable across space. Therefore, even our conservative scenario may be considered optimistic in some cases. In fact, for instance, the wolf is known to disperse in human-dominated landscape crossing highways and agricultural areas (Ciucci et al., 2009), whereas the lynx is known to be much more sensitive to dispersing habitat (Schadt et al., 2002).

Nevertheless, our method is a considerable improvement over the use of simple habitat suitability, because we included functional elements in a structural analysis. This method allows comparing the overall coverage with the effective coverage, which accounts for population structure and species characteristics (Santini et al., 2014). Indeed, networks of PAs are perceived differently by different species (Boitani, 2007). A gap that can be crossed by one species may be perceived as a barrier for more habitat-sensitive species, and a suitable patch for long-term viability for a species may be a mere stepping stone for species with greater area requirements.

Within the range of the three large carnivores, the NPA network covers a much smaller amount of area than the Natura 2000 network, and it almost fully overlaps with the latter. Our results show that the three species of large carnivores are scarcely covered by NPAs. The Natura 2000 network adds considerably to the protection of their habitat; however, most of its sites are too small to support interconnected subpopulations that altogether can form a viable population. Considering population structure, suitable habitat and minimum patch size, only very few countries are able to support viable populations of these species within their PA system.

These results indicate that, even though PAs may contribute to the conservation of large carnivores, they are certainly not sufficient alone. Large carnivores require large areas for their long-term viability and are rather generalists in their habitat selections, and they may tolerate moderately well human disturbance. In general, their conservation requires large-scale connectivity, and more specifically, wide forest cover and abundant prey populations for the lynx and the wolf (Boitani, 2000; Breitenmoser et al., 2000) and trees providing hard mast for the bear (Swenson et al., 2000). A system of PAs such as that provided by the combination of NPA and Natura 2000 sites cannot sustain these species unless the matrix is fully accounted for and managed as part of the overall system (Santini et al., 2014). Most sites of the combined network of PAs may not even provide protection for few individuals. These sites may be important as stepping stones in the dispersing process, as well as core areas for smaller species; however, they are certainly unsuitable for long-term conservation of large carnivores that would require large PAs within a dispersal distance.

Given the insufficient and inadequate coverage given by PAs in Europe, large carnivores should be currently present in only few places and overall declining. Nonetheless, large carnivores in Europe are generally spreading and increasing, and this is likely to be related to matrix intervention (Boitani, 2000; Breitenmoser et al., 2000; Swenson et al., 2000; Linnell et al., 2005; Chapron et al., 2014) rather than to the contribution of the PA system, as most of its sites have been established quite recently (e.g. Natura 2000). PAs may certainly act as refuges and reproduction sites, but it is the matrix that contains most of the population and is crucial to allow long-term viability.

Considering countries in isolation, very few of them can protect one or more viable populations of large carnivores. Therefore, large carnivore conservation needs to be planned at a continental scale. Unfortunately, while Natura 2000 is a European strategy, its planning is at a regional scale, and site selection is at the national level, leading to many uncoordinated national networks (Evans, 2012). Transboundary plans and combined efforts among different nations are required for the conservation of wide-ranging large carnivores. In addition, broad-scale conservation assessments and plans have to account for the species spatial structure in distinct populations (Linnell et al., 2008) and their likely persistence in terms of potential population size. European directives need to integrate recent findings and conservation tools focusing on the whole continent rather than local many uncoordinated efforts.

Acknowledgements

We thank Dr. Paolo Ciucci for the insightful discussions on the topic, Dawid Iurino and Michela Pacifici for comments on the manuscript and Moreno Di Marco and Daniele Baisero for useful suggestions.

References

Allen, C.R., Pearlstine, L.G. & Kitchens, W.M. (2001) Modeling viable mammal populations in gap analyses. *Biological Conservation*, 99, 135–144.

Araùjo, M.B., Lobo, J.M. & Moreno, J.C. (2007) The effectiveness of Iberian protected areas in conserving terrestrial biodiversity. *Conservation Biology*, 21, 1423–1432.

Boitani, L. (2000) *Action plan for the conservation of the wolves (Canis lupus) in Europe*. Council of Europe Publishing, Strasbourg.

Boitani, L. (2007) Ecological networks as conceptual frameworks or operational tools in conservation. *Conservation Biology*, 21, 1414–1422.

Breitenmoser, U., Breitenmoser-Würsten, C., Okarma, H., Kaphegyi, T., Kaphygyi-Wallmann, U. & Müller, U.M. (2000) *Action Plan for the conservation of the Eurasian Lynx (Lynx lynx) in Europe*. Council of Europe Publishing, Strasbourg.

Carbone, C. & Gittleman, J.L. (2002) A common rule for the scaling of carnivore density. *Science*, 295, 2273–2276.

Chapron, G., Kaczensky, P., Linnell, J.D., Von Arx, M., Huber, D., Andrén, H., Lòpez-Bao, J.V., et al. (2014) Recovery of large carnivores in Europe's modern human-dominated landscapes. *Science*, 346(6216), 1517–1519.

Ciucci, P., Reggioni, W., Maiorano, L. & Boitani, L. (2009) Long-distance dispersal of a rescued wolf from the northern Apennines to the western Alps. *The Journal of Wildlife Management*, 73, 1300–1306.

European Commission (2007) Natura 2000 Newsletter No. 21. European Commission, DG ENV, Brussels.

Evans, D. (2012) Building the European Union's Natura 2000 network. *Nature Conservation*, 1, 11–26.

Gaston, K.J., Jackson, S.F., Nagy, A., Cantú-Salazar, L. & Johnson, M. (2008) Protected areas in Europe. *Annals of the New York Academy of Sciences*, 1134, 97–119.

GRASS Development Team (2012) *Geographic resources analysis support system (GRASS) software*. Open Source Geospatial Foundation Project, Michele dll'Adige.

Greenwood, P.J. (1980) Mating systems, philopatry and dispersal in birds and mammals. *Animal Behaviour*, 28, 1140–1162.

IUCN (1994) *Guidelines for protected area management categories*. IUCN Commission on National Parks and Protected Areas, Gland.

Jantke, K., Schleupner, C. & Schneider, U.A. (2011) Gap analysis of European wetland species: priority regions for expanding the Natura 2000 network. *Biodiversity and Conservation*, 20, 581–605.

Jones, K.E., Bielby, J., Cardillo, M., Fritz, S.A., O'Dell, J., Orme, C.D.L., Safi, K., et al. (2009) PanTHERIA: a species-level database of life history, ecology, and geography of extant and recently extinct mammals. *Ecology*, 90, 2648.

Liberg, O. (2006) Genetic aspects of viability in small populations: with special emphasis on the Scandinavian wolf population. Report from an international expert workshop at Färna Herrgård, Sweden, 1–3 May 2002. NATURVÅRDSVERKET.

Linnell, J.D.C., Swenson, J.E. & Andersen, R. (2001) Predators and people: conservation of large carnivores is possible at high human densities if management policy is favourable. *Animal Conservation*, 4, 345–349.

Linnell, J.D.C., Promberger, C., Boitani, L., Swenson, J.E., Breitenmoser, U. & Andersen, R. (2005) The Linkage between Conservation Strategies for Large Carnivores and Biodiversity: The View from the "Half-Full" Forests of Europe. *Large carnivores and the conservation of biodiversity* (ed. Ray J.C., Redford K.H., Steneck, R.S. & Berger J.). Island Press, Washington, DC, p. 381–400.

Linnell, J., Salvatori, V. & Boitani, L. (2008) Guidelines for population level management plans for large carnivores in Europe. A Large Carnivore Initiative for Europe report prepared for the European Commission (contract 070501/2005/424162/MAR/B2).

Maiorano, L., Falcucci, A. & Boitani, L. (2006) Gap analysis of terrestrial vertebrates in Italy: priorities for conservation planning in a human dominated landscape. *Biological Conservation*, 133, 455–473.

Maiorano, L., Falcucci, A., Garton, E.O. & Boitani, L. (2007) Contribution of the Natura 2000 network to biodiversity conservation in Italy. *Conservation Biology*, 21, 1433–1444.

Maiorano, L., Amori, G., Montemaggiori, A., Rondinini, C., Santini, L., Saura, S. & Boitani, L. (2015). On how much biodiversity is covered in Europe by national protected areas and by the Natura 2000 network: insights from terrestrial vertebrates. *Conservation Biology*, 29(4): 986–995.

Natura 2000 (2013). *Natura 2000: the European Network for protected sites.* European Environment Agency, Copenhagen.

Pinto, M.P. & Grelle, C.E.V. (2009) Reserve selection and persistence: complementing the existing Atlantic Forest reserve system. *Biodiversity and Conservation*, 18, 957–968.

R Core Team (2012) *R: A language and environment for statistical computing.* R Foundation for Statistical Computing, Vienna.

Redford, K.H., Coppolillo, P., Sanderson, E.W., Da Fonseca, G.A.B., Dinerstein, E., Groves, C., Mace, G., et al. (2003) Mapping the conservation landscape. *Conservation Biology*, 17, 116–131.

Rondinini, C. & Boitani, L. (2007) Systematic conservation planning and the cost of tackling conservation conflicts with large carnivores in Italy. *Conservation Biology*, 21, 1455–1462.

Rondinini, C. & Pressey, R.L. (2007) Special section: systematic conservation planning in the European landscape: conflicts, environmental changes, and the challenge of countdown 2010. *Conservation Biology*, 21, 1404–1405.

Rondinini, C., Di Marco, M., Chiozza, F., Santulli, G., Baisero, D., Visconti, P., Hoffmann, M., et al. (2011) Global habitat suitability models of terrestrial mammals. *Philosophical Transactions of the Royal Society B*, 366, 2633–2641.

Santini, L., Di Marco, M., Visconti, P., Baisero, D., Boitani, L. & Rondinini, C. (2013) Ecological correlates of dispersal distance in terrestrial mammals. *Hystrix, Italian Journal of Mammalogy*, 24(2), 181–186.

Santini, L., Di Marco, M., Boitani, L., Maiorano, L. & Rondinini, C. (2014). Incorporating spatial population structure in gap analysis reveals inequitable assessments of species protection. *Diversity and Distributions*, 20(6), 698–707.

Schadt, S., Knauer, F., Kaczensky, P., Revilla, E., Wiegand, T. & Trepl, L. (2002) Rule-based assessment of suitable habitat and patch connectivity for the Eurasian lynx. *Ecological Applications*, 12, 1469–1483.

Støen, O.G., Zedrosser, A., Sæbø, S. & Swenson, J.E. (2006) Inversely density-dependent natal dispersal in brown bears Ursus arctos. *Oecologia*, 148, 356–364.

Swenson, J.E., Gerstl, N., Dahle, B. & Zedrosser, A. (2000) *Action plan for the conservation of the brown bear (Ursus arctos) in Europe.* Council of Europe Publishing, Strasbourg.

Treves, A. & Karanth, K.U. (2003) Human-carnivore conflict and perspectives on carnivore management worldwide. *Conservation Biology*, 17, 1491–1499.

Wabakken, P., Sand, H., Kojola, I., Zimmermann, B., Arnemo, J.M., Pedersen, H.C. & Liberg, O. (2007) Multistage, long-range natal dispersal by a global positioning system-collared Scandinavian wolf. *The Journal of Wildlife Management*, 71, 1631–1634.

WDPA (2010). *The world database on protected areas (WDPA).* IUCN & UNEP-WCMC, Gland & Cambridge.

Wielgus, R.B. (2002) Minimum viable population and reserve sizes for naturally regulated grizzly bears in British Columbia. *Biological Conservation*, 106, 381–388.

Woodroffe, R. (2000) Predators and people: using human densities to interpret declines of large carnivores. *Animal Conservation*, 3, 165–173.

Zimmermann, F., Breitenmoser-Würsten, C. & Breitenmoser, U. (2005) Natal dispersal of Eurasian lynx (Lynx lynx) in Switzerland. *Journal of Zoology*, 267, 381–395.

8

Towards Understanding Drivers of Wildlife Population Trends in Terrestrial Protected Areas

M. Barnes[1,2], *I. D. Craigie*[3] *and M. Hockings*[2]

[1]Centre of Excellence for Environmental Decisions, The University of Queensland, St. Lucia, Queensland, Australia
[2]School of Geography Planning and Environmental Management, The University of Queensland, St. Lucia, Queensland, Australia
[3]ARC Centre of Excellence for Coral Reef Studies, James Cook University, Townsville, Queensland, Australia

Introduction

International conservation efforts rely heavily on protected areas, which are commonly considered the most important tool available to maintain habitat integrity and species diversity in the face of increasing anthropogenic threats worldwide (Brooks et al., 2004; Rodrigues, 2006; Andam et al., 2008; Coad et al., 2013). At least 13% of the world's terrestrial land surface area and 2% of its oceans (Spalding et al., 2013) are covered by protected areas. The 2001 Convention on Biological Diversity (CBD) target for 10% coverage by terrestrial protected areas was one of few achieved by the 2010 target date (Coad et al., 2012), and Target 11 of the Aichi protocol calls for the conservation of at least 17% of the world's terrestrial land surface in effectively and equitably managed protected areas by 2020, so they are set to expand further. Protected areas are also a key component of strategies to pursue many of the other Aichi targets, specifically Targets 1, 5, 6, 10, 12, 14, 15 and 17 (CBD, 2010), and underpin the majority of national biodiversity strategies. Protected areas thus represent a massive investment of scarce conservation capital, both directly and in terms of opportunity cost for alternate land uses, including other conservation interventions. Understanding the impact of protected areas on

Protected Areas: Are They Safeguarding Biodiversity?, First Edition.
Edited by Lucas N. Joppa, Jonathan E. M. Baillie and John G. Robinson.

the status and trends of the world's biodiversity is therefore critical in order to evaluate the role that protected areas play in overall conservation strategies.

Despite a steady expansion of the protected area coverage (Bertzky et al., 2012), little progress has been made towards the overarching goal of halting the decline of global biodiversity (Butchart et al., 2010). Biodiversity threats are increasing (Butchart et al., 2010; WWF, 2012, 2014), including within and around many protected areas (Laurance et al., 2012). As a result, the efficacy of protected areas has been challenged (Boitani et al., 2008; McDonald-Madden et al., 2011). The argument for the creation of protected areas to date has been based on the rationale that either declaration per se or in situ management will permit the site to maintain or improve biodiversity outcomes. Unfortunately, simply declaring an area protected is insufficient to guarantee biodiversity conservation, especially if management is ineffective. Quantitative evaluation of wildlife outcomes in protected areas has been relatively uncommon to date (Geldmann et al., 2013), and protected area performance in maintaining biodiversity values remains poorly understood, in particular for wildlife species. Unless the impact of protected areas on conservation outcomes and the circumstances under which they are successful are clarified, the effectiveness of protected areas in safeguarding biodiversity will continue to be called into question, and rightfully so. Continued support and investment in protected areas demands the demonstration that they are indeed effective for wildlife species, at least when managed appropriately.

High-quality wildlife monitoring data are rare (Western et al., 2009; Eaton et al., 2012). As a result, limited data are available with which to evaluate population condition or status of species of concern. Population trends in protected areas are rarely known, and even where they are, the causes of change are not always apparent. Available data indicate that wildlife outcomes in protected areas are highly variable (Craigie et al., 2010) and that the relationships between trends and their causes of change are highly complex. Possible drivers are often highly co-correlated (e.g. International Union for Conservation of Nature (IUCN) protected area category (strictness of protection) and protected area size), and understanding is hampered by a lack of appropriate and coordinated assessment. Even some of what might be considered to be the best monitored populations lack sufficient data to conclusively determine the effectiveness of specific management interventions (Walsh et al., 2012).

Some of the most useful data to test and explore protected area effectiveness are population time series where the abundances of key species in protected areas have been measured over a number of years, using a method that allows direct comparison between years. Ideally, such monitoring would be conducted in an adaptive management framework, with observed changes in populations guiding future management priorities. However, this is rarely the case for several reasons; budgets are limited and managers are often reluctant to divert scarce resources away from management actions to monitoring and evaluation (Kapos et al., 2008; McDonald-Madden et al., 2011), so monitoring is often neglected. Even when monitoring data are collected by management agencies, they are rarely published or peer reviewed making it generally difficult to access in order to conduct synthesis or other comparative analysis. Additionally, institutional data can have a tendency to be lost, for example, Parks Canada conservatively estimated that 50% of their ecological data was lost over a 20-year period, citing poor documentation of metadata, changing data standards and formats and poor access to grey literature (Parks Canada, 2007). Some time series data are collected for purposes other than management (e.g. ecological or behavioural research conducted by universities and non-governmental organisations (NGOs)), and these data are more often in the public domain (Howe & Milner-Gulland, 2012).

In addition, data on the drivers of change of wildlife abundances is often unavailable and is thus a constraint on the analyses that are possible. In the locations where time series are available, information regarding resources, inputs, management actions and contextual factors are often unavailable. In reality, for management agencies, considerable, often unachievable, levels of investment in research are necessary to evaluate drivers of protected area effectiveness at broad scales (Howe & Milner-Gulland, 2012). In the absence of evidence, decisions about management actions and protected area planning and implementation are informed by experience, intuition or past practice at best and anecdote at worst (Cook, 2010; Bottrill et al., 2011). Recent data suggest that up to 90% of conservation assessments are made without evidence-based knowledge (Sutherland et al., 2004; Cook et al., 2010).

Improved understanding of drivers would lead to management and policy that are more likely to maximise species outcomes in protected areas, and this would yield tangible conservation benefits in the short and long term. In this chapter, we therefore present findings from evaluations of recent and historical literature and examine three new lines of evidence that illuminate possible drivers of wildlife population outcomes in terrestrial protected areas. First, we examine emergent factors as identified in the literature, examine the findings of the global study on management effectiveness and present two contrasting quantitative case studies from Canada and Africa. We highlight key synergies and differences between the findings of these studies.

Impact of protection

The findings of a recent systematic review (Geldmann et al., 2013) highlight major deficiencies in our ability to link changes in inputs, actions and context to changes in biodiversity values (biological outcomes), especially for wildlife species. The review focuses explicitly on studies where quantitative impact evaluation has been conducted and highlights the limited availability of evidence on the impact of protected areas on habitats and fauna. Alarmingly, there is very little quantitative understanding of how, and under what conditions, different protected area management interventions improve protected area effectiveness for wildlife populations; only 42 studies that included appropriate counterfactuals for species populations were located by the systematic review, and the overall effect of protection is unclear. Where impact evaluation occurs, it is usually at a limited geographic scale and synthesis is rare (Butchart et al., 2010; Hoffmann et al., 2010; Geldmann et al., 2013). Consequently, understanding of the conservation impact of protected areas on wildlife populations globally remains limited.

Other attempts to evaluate the impact of protected areas on species conservation outcomes (both in and outside of protected areas) have relied on more indirect techniques such as expert interviews (Bruner et al., 2001; Hoffmann et al., 2010; Laurance et al., 2012) or used aggregate indices like the Living Planet Index (Loh et al., 2005; Collen et al., 2009). However, only some of these have been applied to protected areas and even fewer address outcomes in protected areas systematically (Craigie et al., 2010; Laurance et al., 2012). Hoffmann et al. (2010) conservatively estimated that the rate of deterioration in threatened status among vertebrates would be at least 20% worse in the absence of all conservation measures, including protected areas, but suggested that current conservation efforts remain insufficient to offset the main drivers of biodiversity loss in this group. To halt the decline of biodiversity in protected areas, the next step is to identify the types of actions, attributes and contexts that impact biodiversity outcomes and invest in those that correspond with positive impacts.

Drivers of biodiversity outcomes

Despite the dearth of impact evaluation, it is possible to identify a suite of probable and possible factors likely to influence biodiversity outcomes in protected areas, based on both theoretical and empirical studies (Barnes, 2013). Both 'in' and 'ex situ' factors are important, and they can be broadly grouped into the themes of design (e.g. size, shape), management (e.g. resources, visitation), sociopolitical context (e.g. statutory protection, corruption), ecological context (e.g. sensitivity to disturbance, fecundity) and threat context (e.g. poverty, land use change). However, relationships between these factors and positive biodiversity outcomes appear complex and interrelated, and often, the factors thought to be important lack explanatory power (Bruner et al., 2001; Hilborn et al., 2006; Leverington et al., 2008; Craigie, 2011).

Management effectiveness evaluation

One tool that has been used to evaluate the link between site-specific management actions, context and outcomes is management effectiveness evaluation (MEE). For instance, Hockings et al. (2004) identified ineffective and weak management as a key reason for ongoing loss of biodiversity values in protected areas at the World Parks Congress in Durban, 2003. To address this problem, MEE aims to evaluate all aspects of a protected area management, from technical capacity to visitor management and biodiversity targets (Ervin, 2003; EarthTrends, 2010; WDPA, 2010) across the management cycle, including design, adequacy and delivery of objectives in order to determine the extent to which management is being implemented and where possible whether it is protecting identified values and achieving goals and objectives (Hockings et al., 2004). MEE relies largely on expert judgements of management effectiveness using one of the numerous systems developed under the IUCN management effectiveness framework (Hockings et al., 2004) and is designed to highlight deficiencies so that they can be addressed and improved.

Evaluation of management effectiveness is a vital component of adaptive management of protected areas. In recognition of the capacity of evaluation to improve management, and in doing so potentially improve conservation outcomes in protected areas, the CBD Conference of the Parties (COP) 7 adopted the Programme of Work on Protected Areas (PoWPA) that called on member nations to evaluate management effectiveness in at least 30% of their protected areas by 2010 (CBD, 2004; Hockings et al., 2004) and revised this upwards to 60% of protected areas by 2015 in COP 10 decision X31 (Woodley et al., 2012). Since then, the number of protected areas subjected to MEEs has increased rapidly (CBD, 2010; Leverington et al., 2010) with thousands of site assessments that have now been conducted globally. However, the extent of data available varies widely across sites (Leverington et al., 2010), and while the guidelines for evaluation prepared by IUCN (Hockings et al., 2006) encourage the use of quantitative data to inform the evaluation where possible, the reality in many protected areas is that such data are frequently not available. Although the 30% target was not achieved, positive changes have been implemented in many protected areas and protected area networks in response to status evaluations (Hockings et al., 2009).

Many MEE systems only assess achievement of outcomes at a broad level (Leverington et al., 2010), which makes it difficult to link individual actions or management strategies to any biodiversity outcomes, particularly given the complexity of the relevant socioecological systems. As a result, input and process measures often have a poorly quantified relationship to specific biodiversity outcomes, limiting their utility for understanding potential drivers of conservation results at a site. Rather, such evaluation systems seek to identify and relate broad outcomes of management to the adequacy and appropriateness of management systems and processes.

In a global study on management effectiveness, Leverington et al. (2010) compiled and examined data from 1968 assessments of management effectiveness in 95 countries, to determine key indicators of performance. Management factors that strongly correlated with positive condition of natural values were:

- Skills of staff and other management partners
- Effective natural resource management (NRM) programmes
- Research and monitoring

- Constraint or support by the external civil and political environment
- Achievements of work programmes
- Adequacy of law enforcement

However, it is not possible to disaggregate species, habitat and other biological outcomes (Leverington et al., 2008, 2010), and counterfactual data are not available so it is not possible to establish causative links between these factors and the biodiversity outcomes in the sites.

Kelman (2010) identified similar relationships between inputs and outcomes in New South Wales (NSW), Australia,[1] which thus supported the conclusions of the global assessment. In NSW, relative planning and staff time allocation and landscape context were identified as the primary factors influencing management effectiveness scores in NSW protected areas. Positive results for threatened species in NSW parks were strongly linked with the existence of and implementation of actions from species-specific management plans. Kelman noted that more than one-quarter of protected areas in her sample lacked sufficient information for effective management of threatened species and that parks with poor overall management effectiveness outcomes scores were more likely to lack sufficient species monitoring data. Pattern analysis also highlighted a strong relationship between resource sufficiency and positive outcomes; where scaling of resources in response to protected area size and pressures such as visitation occurred, biological outcomes were more positive (Kelman, 2010).

Both the global and NSW studies suggested that management inputs can successfully mitigate threats and highlighted probable links between resourcing and planning and positive conservation outcomes. However, MEEs are not primarily designed to focus on detailed understanding of biodiversity outcomes in sites and more detailed studies, which underpinned that good biodiversity monitoring data are required.

Case study: Canada

Canada has a strong history of species monitoring and high-resolution spatial data for a variety of landscape and socio-economic attribute data across the country. Parks Canada has comprehensive management planning, a history of high-quality species-based monitoring and a good understanding and documentation of historical resourcing and pressures (Parks Canada, 2007), and the Canadian protected areas system is under increasing pressure from a variety of anthropogenic threats (Kramer & Doran, 2010). The data-rich environment in Canada combined with the rare willingness to share institutional data makes the Canadian Federal Reserve System ideal as a case study to identify potential drivers of biodiversity outcomes in protected areas.

We compiled a database of all available population abundance time series for Canadian native vertebrates within protected areas using data from a variety of sources, including the Parks Canada ICE database, Environment Canada, published literature and the Living Planet Index database. Species monitoring in Canada focuses predominantly on mammals, such as caribou (*Rangifer tarandus*), bison (*Bison bison*), beavers (*Castor canadensis*) and birds such as sandhill cranes (*Grus canadensis*), harlequin ducks (*Histrionicus histrionicus*) and olive-sided flycatchers (*Contopus cooperi*); therefore, other taxonomic groups have not been examined. Time series exhibit a large range of variability in species outcomes and considerable uncertainty, which reflects both the underlying variability in the systems and uncertainty surrounding sampling and census methods.

Monitored vertebrates in Canadian protected areas appear to have exhibited early declines (see also Rivard et al., 2000), but more recently, Canadian protected areas have been successful in maintaining much of their vertebrate fauna, in stark contrast to ongoing global declines (Butchart et al., 2010). Canadian national parks also appear to be doing slightly better when compared to temperate wildlife population trends in general (Fergusson et al., 2012; WWF, 2012), which are better than the global average (WWF, 2012).

We examined a suite of explanatory variables likely[2] to influence species population outcomes across 21 protected areas in the Canadian Federal Parks Estate, where detailed visitation, staff, budget and expenditure data was available (Barnes, 2013). Insight from Canada can be applied to adapt and optimise protected area management policies to maximise species conservation outcomes both in Canada and worldwide. Linear mixed effects models (LMEs) were applied to explore variation across sites and species[3] and identify key correlates of performance for 185 populations in 21 protected areas. Modelling indicated a mixture of management inputs and species, and site-level ecological characteristics were the best predictors of population trends. In particular, species threat status,[4] capital expenditure,[5] staff time,[6] age of the protected area and species body mass were the most important variables for predicting biodiversity outcomes. Size (area) of protected areas and latitude were also weakly important (Barnes, 2013).

Capital expenditure and staff time both exhibit the expected trend – greater investment correlates with more positive outcomes. If sufficient infrastructure and staff are available, protected areas have the capacity to mitigate threatening processes. Legal mechanisms tend to be more effective in wealthy countries (Barett et al., 2001; Wright et al., 2007), so it is uncertain to what extent these conclusions hold true in less wealthy contexts. The importance of capital expenditure (and staff time) reiterates the need for sufficient resources and staff dedicated to NRM, including monitoring and evaluation. If sufficient infrastructure and equipment (capital expenditure) and staff are available, protected areas have the capacity to mitigate threatening processes (e.g. Horsup, 2004; Timko & Innes, 2009). Spending across all of Canada's protected areas is unknown, but approximately US$1 billion per annum was spent on all biodiversity conservation in Canada (between 2001 and 2007) including federal and provincial government funding to parks, the Northern Biodiversity Fund, user fees and agri-environmental funds from government statistical summaries (Waldron et al., 2013). These findings illustrate that in Canada, a country with generally strong environmental regulation and management (Esty, 2006), at least until recently (Favaro et al., 2012), when substantial capital resources and time are invested in protected area management, biodiversity values can be maintained and improved.

Case study: Sub-Saharan Africa

Sub-Saharan Africa has a number of characteristics that make it a worthwhile place to study protected area performance. It has a relatively high proportion of its land surface protected (Coad et al., 2012) and it has a long history of protected area establishment dating back to the 19th century (Cumming, 2004). It contains a wealth of biodiversity (Myers et al., 2000; Shorrocks, 2007), much of which is under severe pressure from poverty, development and human population growth (United Nations, 2001; Cordeiro et al., 2007), and studies have shown that there is a wide variation in the conservation performance of protected areas across the region as measured by vertebrate population trends (Brashares et al., 2001; Whyte, 2005; Craigie et al., 2010). In an effort to explore the causes of the variation in population trends between sites and species, LMEs were used to examine 586 population time series from 81 protected areas. We collated time series information on the population sizes of mammal species in continental African protected areas from published scientific literature, online databases (SANParks, 2009; NERC Centre for Population Biology, 2010), park management reports, the Living Planet Index database (McRae et al., 2012) and personal communications of unpublished data. The full details and overall patterns in these data can be found in Craigie et al. (2010), but in general, there was a substantial decline (~59%) in the overall population abundances of these species, which were predominantly (>90%) large (>5 kg) herbivorous mammals.

An ordinary least squares regression was fitted to each log-transformed individual population time series and the slope value of the regression used as the dependent variable. A wide range of potential explanatory variables was evaluated through modelling; these included the characteristics of the protected areas (such as their size, surrounding human population density and staffing levels) and the species traits of the vertebrates they contain. The results showed that species trends were more positive in protected areas with more staff, where hunting was not a major threat, for larger-bodied species, and for time series dated after around 1990. The influence of protected area size was complex, with some of the largest sites showing especially poor outcomes. Results show that larger protected areas performed more poorly than smaller protected areas overall, with the likely cause being disproportionally fewer staff per unit area in larger protected areas. These results show that an increased level of resources is likely to lead to improved conservation performance and that resource levels can have more influence on outcomes than other ecologically important features such as protected area size. Further, more recent increases illustrate the pay-off from aggressive anti-poaching measures, but unfortunately, the data here do not reflect recent declines in the face of overwhelming wildlife crime, which has reversed some of these increases.

These results from sub-Saharan Africa support the findings of other studies that show the size of budgets and numbers of staff for protected areas that are consistently important correlates of protected area performance (Leader-Williams & Albon, 1988; Dudley et al., 2004). Larger protected area budgets, staff numbers and equipment levels allow protected areas to better mitigate the threats facing the biodiversity they contain (Jachmann, 2004, 2008). The mechanisms that this mitigation can take are often specific to the threat; in many African protected areas, armed patrols are used to deter poachers (Caro et al., 1998), and the effectiveness of anti-poaching activity has been shown to be related to the intensity of effort (Milner-Gulland & Leader-Williams, 1992; Leader-Williams & Milner-Gulland, 1993; Hilborn et al., 2006). In these data, it appears that the effect of resource shortages is outweighing the likely ecological beneficial effect of larger protected areas. The practical implication of this is that funding bodies need to target larger protected areas with additional resources to enable them to perform adequately and maintain their wildlife populations.

In a similar result to the Canadian case study, larger body mass is correlated with more positive population trends. This leads to the question, why should larger-bodied species be faring better than smaller ones in African protected areas? One possibility is that protected area staffs are focusing their limited resources on them, perhaps because they are preferred by fee-paying tourists (Hilborn et al., 2006). Alternatively, bigger-bodied species are more readily monitored (or found as carcasses), so threats to them may be more quickly addressed. A third possibility is that poachers preferentially target smaller-bodied species because this is less dangerous, it requires cheaper equipment (de Merode et al., 2004; Setsaas et al., 2007), it incurs a lower risk of detection, capture or punishment (Milner-Gulland & Leader-Williams, 1992) or because smaller carcasses are easier to transport out of the protected area. The implication of this is that conservation efforts may need to refocus on these small- and medium-sized species that are not endangered today but may become so tomorrow.

Synthesis

Quantifying what makes protected areas effective is an important challenge for biodiversity conservation. A critical constraint in addressing this question is that to date it has proven difficult to carry out analyses which can be extrapolated to inform conservation management widely, rather being solely informative for local management (i.e. case studies such as Hilborn et al., 2006; Armsworth et al., 2011). Nevertheless, we identified a suite of factors that available data indicate are likely to be important in determining wildlife outcomes in protected areas. We have focused primarily on management, threat context and design characteristics in the case studies and will discuss emergent findings from those studies here.

The evidence examined in this chapter suggests that the effectiveness of protected areas in achieving wildlife outcomes is limited, at least in part, by resources, both in terms of money and capacity (which are often interlinked). Notably, across the strands of evidence examined, indicators of capacity are consistently important in predicting improvement in wildlife populations. Although design factors were important in some cases, management factors and, in particular, appropriate resourcing and capacity were consistently strongly related to trends, often outweighing the explanatory power of other ecological and contextual factors.

In the global study of management effectiveness, management skills of staff and management partners, whether planned work programmes were delivered and whether protected areas are legally designated, were important (Leverington et al., 2010). In the Canadian case study, greater resources (specifically capital expenditure and staff time) were found to be the most important variables for predicting more positive population trends (Barnes, 2013). Similarly in sub-Saharan Africa, results showed that species trends were more positive in protected areas with more staff. And in the systematic review, investment in anti-poaching (albeit in a small sample) resulted in more positive outcomes in protected areas – though this wasn't always enough to halt declines. Promisingly, together these findings suggest that the tools for achieving desired biodiversity outcomes exist, if we can leverage the resources to do so.

Resources are linked to capacity to enforce regulations and to enact specific management actions (Leverington et al., 2010), and these studies add substantial support to other recent findings to this effect (Chapter 11). For instance, Walston et al. (2010) demonstrated that a minimum of US$930 per km^2 per year is required in order to retain tiger populations in Southeast Asia. Sufficient and sustainable financing of management is therefore critical. If sufficient infrastructure, equipment (capital expenditure) and staff are available, protected areas have the capacity to mitigate threatening processes (Horsup, 2004; Timko & Innes, 2009; Craigie, 2011; Barnes, 2013). Shortfalls in resources, including staff time and capacity, equipment, capital infrastructure and operating and project funds, have been shown to result in suboptimal biodiversity outcomes in protected areas (sensu Hilborn et al., 2006). It is probable that the demonstrated relationship between resource and management capacity factors and biological outcomes is related to the ability of managers to do work to mitigate threatening processes. For instance, clearly identified and enforced boundaries are commonly reported as critical management actions in the face of direct threats (Hilborn et al., 2006; Leverington et al., 2010; Geldmann et al., 2013), and sites with management plans generally score more highly in MEEs than those without (Leverington et al., 2010).

Although theory predicts protected area size to be extremely important in determining the capacity of a protected area to retain species, and certainly habitat area predicts the carrying capacity of landscapes (Newmark, 1986, 1987, 1996), the influence of the size of protected areas is repeatedly outweighed by management factors in the case studies examined. In Canada, protected area size was only weakly important, and in other recent global evaluations (Laurance et al., 2012), it also had little explanatory power. These findings suggest that in reality, a more complex context exists where outcomes are often more strongly influenced by other factors. In sub-Saharan Africa, the effect of protected area size was more complex, and perhaps counter-intuitively, larger protected areas performed more poorly than smaller protected areas. The likely reason for this was identified as being disproportionately fewer staff per unit area in larger protected areas, which interacted with the influence of area in a way that reversed the predicted effect when both were included in the model, suggesting that increased resources may lead to improved conservation performance in the region, which would mirror findings for megafauna in Southeast Asia (Walston et al., 2010).

Protected area size is commonly cited as being of critical importance, but we find only minor empirical support for its influence. It is likely that decadal-scale processes make it difficult to detect the influence of size. It is however not protected area size per se but rather effective habitat size that makes protected area size important (Newmark, 1986, 1987), and this is much more difficult to ascertain. Some protected area may be listed as very small but embedded within a larger reasonably intact ecosystem and therefore protected de facto and functionally much larger, while others may be reasonably large but isolated and therefore functionally much smaller (DeFries et al., 2005). Functional extent is also likely to depend upon the degree of threat in the greater functional landscape and is both species and habitat specific; for instance, in protected areas with multiple habitat types, a particular habitat may comprise only a fraction of the total area. However, it may also be that active management, if resourced and implemented appropriately, has more influence on outcomes than other ecologically important features such as protected area size, suggesting that well-managed protected areas, at least, may be capable of sustaining wildlife populations in the face of ongoing decline.

Undoubtedly, other factors are important in understanding biological outcomes in protected areas, but they may be unmeasured or measured at inappropriate times or scales, or they may be outweighed by management effort.

Concluding remarks

Triangulation of a variety of recent research indicates that effective management and sufficient resourcing to conduct management activities are clearly important for retaining conservation values and that effective management can outweigh threats, at least in well-resourced systems. In light of the importance of effective management, the current focus of research into ecological factors affecting populations might be beneficially redirected to more applied research which focuses on improving protected area management, since management improvements appear likely to achieve the greatest impact on biodiversity benefits.

Of note, however, is that the studies available are generally biased towards well-resourced systems (e.g. Canada) and protected areas, even within systems where resourcing is variable (e.g. sub-Saharan Africa), as it is here that population monitoring has been conducted. It would be valuable to undertake similar regional assessments to the Canadian and sub-Saharan case studies across wider variety of socio-economic contexts and resourcing contexts. To achieve this, we will require ongoing evaluation of outcomes and well-designed monitoring programmes that specifically link (and monitor) actions, investments and outputs as well as the outcomes themselves. The most fruitful research would include counterfactual controls and a sample that included the full spectrum of resourcing.

High costs associated with good-quality monitoring over sufficient time frames often make it prohibitively expensive to produce useful population data (Howe & Milner-Gulland, 2012). However, the most wasteful action possible is to spend resources on protected areas that achieve minimal biodiversity benefit. It is therefore imperative to measure outcomes to avoid this lose-lose scenario. Programme guidelines can help to redress this issue by dedicating a minimum proportion of the budget for outcomes-based monitoring and targeting data collection that can add a large amount of value quickly, for example, by substantially reducing uncertainty or improving power (Hauser et al., 2006). Creative solutions, such as more widespread and proactive utilisation of citizen science data, will also be necessary to provide information to facilitate decision-making within a resource limited context.

It is possible to achieve effective conservation in protected areas, but fundamentally, this depends upon appropriate implementation and funding of protected areas worldwide, and hard decisions will need to be made; as given the limited resources and capacity for conservation, it will be impossible to make all protected areas effective in conserving all species of wildlife.

Acknowledgements

We would like to acknowledge the members of the IUCN Task Force on Biodiversity and Protected Areas for their valuable intellectual support; Jonathan Loh, Louise McRae, WWF and ZSL for access to Living Planet Index data and code and Ben Collen for related analytical support; Parks Canada, Stephen Woodley and their dedicated staff for site data and Paul Zorn and Dave Hayes for spatial data and technical discussions. We acknowledge the IUCN Working Group on management effectiveness whose work forms the basis of one section of this chapter. We also acknowledge the work of our co-authors on the systematic review, Jonas Geldmann, Neil Burgess and Lauren Coad. Josie Kelman also provided comments.

Notes

1. The NSW National Parks and Wildlife Service (NPWS) State of the Parks (SOP) assessment is one of the most comprehensive network scale management effectiveness assessments ever conducted.

2. 'Likely' explanatory variables were selected based on literature review and available data sets.
3. Species, Site and Management Unit were included as random effects.
4. IUCN Red List Status.
5. Capital expenditure is defined by Parks Canada as 'major expenditures for, and major repairs to, infrastructure and significant equipment – roads, buildings, trucks, etc. It does not cover regular maintenance'.
6. Staff time is defined as full-time equivalent person years spent on natural resource management or related roles, based on the mean salary of resource conservation staff across Parks Canada.

References

Andam, K. S. et al., 2008. Measuring the effectiveness of protected area networks in reducing deforestation. *Proceedings of the National Academy of Sciences of the United States of America*, 105(42), p. 16089–16094.

Armsworth, P. R. et al., 2011. Management costs for small protected areas and economies of scale in habitat conservation. *Biological Conservation*, 144(1), p. 423–429.

Barett, C. B., Brandon, K., Gibson, C. & Gjertsen, H., 2001. Conserving tropical biodiversity amid weak institutions. *BioScience*, 51(6), p. 497–502.

Barnes, M., 2013. *Protected area effectiveness: Evaluation of biological outcomes in protected areas*, Brisbane: School of Geography, Planning and Environmental Management, University of Queensland.

Bertzky, B. et al., 2012. *Protected planet report 2012: Tracking progress towards global targets for protected areas*, Gland & Cambridge: IUCN & UNEP-WCMC.

Boitani, L. et al., 2008. Change the IUCN protected area categories to reflect biodiversity outcomes, *PLoS Biology*, 6, p. 436–438.

Bottrill, M. C., Hockings, M. & Possingham, H. P., 2011. In pursuit of knowledge: Addressing barriers to effective conservation evaluation. *Ecology and Society*, 16(2), p. 14.

Brashares, J. S., Arcese, P. & Sam, M. K., 2001. Human demography and reserve size predict wildlife extinction in West Africa. *Proceedings of the Royal Society of London Series B*, 268(1484), p. 2473–2478.

Brooks, T. M. et al., 2004. Coverage provided by the global protected area system: Is it enough. *BioScience*, 54(12), p. 1081–1091.

Bruner, A. G., Gullison, R. E., Rice, R. E. & da Fonseca, G. A. B., 2001. Effectiveness of parks in protecting tropical biodiversity. *Science*, 291(125), p. 125–128.

Butchart, S. H. M. et al., 2010. Global biodiversity: Indicators of recent declines. *Science*, 328(5982), p. 1164–1168.

Caro, T. M. et al., 1998. Consequences of different forms of conservation for large mammals in Tanzania: Preliminary analyses. *African Journal of Ecology*, 36, p. 303–320.

CBD, 2004. *Decisions adopted by the CoP to the CBD at its 7th meeting (UNEP/CBD/COP/7/21)*, Montreal: Secretariat of the CBD.

CBD, 2010. *Decisions adopted by the CoP to the CBD at its 10th meeting (UNEP/CBD/COP/DEC/X/2)*, Montreal: Secretariat of the CBD.

Coad, L., Burgess, N. D., Bombard, B. & Besancon, C., 2012. *Progress towards the CBD's 2010 and 2012 targets for protected area coverage*, Cambridge: UNEP-WCMC.

Coad, L. et al., 2013. Progression towards the CBD protected area management effectiveness targets. *PARKS*, 19(1), p. 13–24.

Collen, B. et al., 2009. Monitoring change in vertebrate abundance: The living planet index. *Conservation Biology*, 23, p. 317–327.

Cook, C., 2010. *Validating the use of expert opinion in management effectiveness assessments of protected areas in Australia*, Brisbane: School of Integrative Systems, University of Queensland.

Cook, C. N., Hockings, M. & Carter, R., 2010. Conservation in the dark? The information used to support management decisions. *Frontiers in Ecology and the Environment*, 8(4), p. 181–186.

Cordeiro, N. J. et al., 2007. Conservation in areas of high population density in Sub-Saharan Africa. *Biological Conservation*, 134(2), p. 155–163.

Craigie, I. D., 2011. *An assessment of the performance of Africa's protected areas*, Cambridge: Department of Zoology, University of Cambridge.

Craigie, I. D. et al., 2010. Large mammal population declines in Africa's protected areas. *Biological Conservation*, 143, p. 2221–2228.

Cumming, D., 2004. Performance of parks in a century of change. In: B. Child, ed. *Parks in transition: Biodiversity, rural development and the bottom line*. London: Earthscan, p. 105–125.

DeFries, R., Hansen, A. J., Newton, A. & Hansen, M. C., 2005. Isolation of protected areas in tropical forests over the last twenty years. *Ecological Applications*, 15(1), p. 19–26.

Dudley, N. et al., 2004. *Are protected areas working*, Gland: WWF.

EarthTrends, 2010. Earthtrends Environmental Information. [Online] Available at: http://earthtrends.wri.org [Accessed 13 July 2015].

Eaton, M. A. et al., 2012. *The state of the UK's birds 2012*, Sandy: RSPB.

Ervin, J., 2003. Protected area assessment in perspectives. *BioScience*, 53(9), p. 819–822.

Esty, D. C., 2006. *Environmental performance index: The fate of the earth, a roundtable on environmental assessment*, New Haven: American Enterprise Institute.

Favaro, B., Reynolds, J. D. & Cote, I. M., 2012. Canada's weakening aquatic protection. *Science*, 337(6091), p. 154.

Fergusson, S. H. et al., 2012. Time series data for Canadian arctic vertebrates: IPY contributions to science, management and policy. *Climatic Change*, 115(1), p. 235–358.

Geldmann, J. et al., 2013. Effectiveness of terrestrial protected areas in reducing habitat loss and population declines. *Biological Conservation*, 161, p. 230–238.

Hauser, C. E., Pople, A. R. & Possingham, H. P., 2006. Should managed populations be monitored every year. *Ecological Applications*, 16(2), p. 807–819.

Hilborn, R. et al., 2006. Effective enforcement in a conservation area. *Science*, 314(5803), p. 1266.

Hockings, M., Stolton, S. & Dudley, N., 2004. Management effectiveness: Assessing management of protected areas. *Journal of Environmental Policy and Planning*, 6(2), p. 157–174.

Hockings, M. et al., 2006. *Evaluating effectiveness: A framework for assessing management effectiveness of protected areas*. 2nd ed. Gland: IUCN.

Hockings, M., Cook, C., Carter, R. & James, R., 2009. Accountability, reporting or management improvement? Development of a state of the parks assessment system in New South Wales, Australia. *Environmental Management*, 43(6), p. 1013–1025.

Hoffmann, M. et al., 2010. The impacts of conservation on the status of the world's vertebrates. *Science*, 330(6010), p. 1503–1509.

Horsup, A., 2004. *Recovery plan for the northern hairy-nosed wombat (Lasiorhinus Krefftii) 2004–2008*, Brisbane: State of Queensland EPA.

Howe, C. & Milner-Gulland, E. J., 2012. Evaluating indices of conservation success: A comparative analysis of outcome and output based indices. *Animal Conservation*, 15(3), p. 217–226.

Jachmann, H., 2004. Monitoring law enforcement performance in nine protected areas in Ghana. *Biological Conservation*, 141(1), p. 89–99.

Jachmann, H., 2008. Illegal wildlife use and protected area management in Ghana. *Biological Conservation*, 141(7), p. 1906–1918.

Kapos, V. et al., 2008. Calibrating conservation: New tools for measuring success. *Conservation Letters*, 1(4), p. 155–164.

Kelman, J., 2010. *Management effectiveness in NSW*, Brisbane: School of Integrated Systems, University of Queensland.

Kramer, D. B. & Doran, P. J., 2010. Land conversion at the protected area's edge. *Conservation Letters*, 3(5), p. 349–358.

Laurance, W. F. et al., 2012. Averting biodiversity collapse in tropical forest protected areas. *Nature*, 489, p. 290–294.

Leader-Williams, N. & Albon, S. D., 1988. Allocation of resources for conservation. *Nature*, 336, p. 533–535.

Leader-Williams, N. & Milner-Gulland, E. J., 1993. Policies for the enforcement of wildlife laws: The balance between detection and penalties in Luangwa Valley, Zambia. *Conservation Biology*, 7(3), p. 611–617.

Leverington, F., Hockings, M. & Lemos-Costa, K., 2008. *Management effectiveness evaluation in protected areas*, Gatton: University of Queensland.

Leverington, F. et al., 2010. A global analysis of protected area management effectiveness. *Environmental Management*, 46(5), p. 685–698.

Loh, J. et al., 2005. The living planet index: Using species population time series to track trends in biodiversity. *Philosophical Transactions of the Royal Society of London Series B*, 360(1454), p. 289–295.

McDonald-Madden, E. et al., 2011. Should we implement monitoring or research for conservation. *Trends in Ecology and Evolution*, 26(3), p. 108–109.

Living Planet Index, 2012. The living planet index data report. London: ZSL & WWF.

de Merode, E., Homewood, K. & Cowlishaw, G., 2004. The value of bushmeat and other wild foods to rural households living in extreme poverty in the Democratic Republic of Congo. *Biological Conservation*, 118(5), p. 573–581.

Milner-Gulland, E. J. & Leader-Williams, N., 1992. A model of incentives for the illegal exploitation of black rhinos and elephants: Poaching pays in Luangwa Valley, Zambia. *Journal of Applied Ecology*, 29(2), p. 388–401.

Myers, N. et al., 2000. Biodiversity hotspots for conservation priorities. *Nature*, 403(6772), p. 853–858.

NERC Centre for Population Biology, 2010. Dynamics Database Version 2. [Online] Available at: http://www.sw.ic.ac.uk/cpb/gpdd.html [Accessed 13 July 2015].

Newmark, W. D., 1986. Species-area relationship and its determinates for mammals in western North American national parks. *Biological Journal of the Linnean Society*, 28(1–2), p. 83–98.

Newmark, W. D., 1987. A land-bridge island perspective on mammalian extinctions in western North American parks. *Nature*, 325(6103), p. 430–432.

Newmark, W. D., 1996. Insularization of Tanzanian parks and the local extinction of large mammals. *Conservation Biology*, 10(6), p. 1549–1556.

Parks Canada, 2007. *Monitoring and reporting ecological integrity in Canada's national parks: Volume 2 – A park level guide to establishing EI monitoring*, Gatineau: Parks Canada.

Rivard, D. H. et al., 2000. Changing species richness and composition in Canadian national parks. *Conservation Biology*, 14(4), p. 1099–1109.

Rodrigues, A. S. L., 2006. Are global conservation efforts successful. *Science*, 313, p. 1051–1052.

SANParks, 2009. South African National Parks Data Repository. [Online] Available at: http://dataknp.sanparks.org/sanparks [Accessed 3 March 2009].

Setsaas, T. H. et al., 2007. How does human exploitation affect impala populations in protected and partially protected areas? A case study from the Serengeti Ecosystem. *Biological Conservation*, 136, p. 563–570.

Shorrocks, B., 2007. *The biology of African savannahs*, Oxford: Oxford University Press.

Spalding, M. D. et al., 2013. Protecting marine spaces: Global targets and changing approaches. *Ocean Yearbook*, 27, p. 213–248.

Sutherland, W. J., Pullin, A. S., Dolman, P. M. & Knight, T. M., 2004. The need for evidence based conservation. *Trends in Ecology and Evolution*, 19(6), p. 305–308.

Timko, J. A. & Innes, J. L., 2009. Evaluating ecological integrity in national parks: Case studies from Canada and South Africa. *Biological Conservation*, 142(3), p. 676–688.

United Nations, 2001. *World population monitoring 2001: Population, environment and development*, New York: United Nations.

Waldron, A. et al., 2013. Targeting global conservation funding to limit immediate biodiversity declines. *Proceedings of the National Academy of Sciences of the United States of America*, 110(29), p. 12144–12148.

Walsh, J. C., Wilson, K. A., Benshemesh, J. & Possingham, H. P., 2012. Unexpected outcomes of invasive predator control: The importance of evaluating conservation management actions. *Animal Conservation*, 15(4), p. 319–328.

Walston, J. et al., 2010. Bringing the tiger back from the brink: The six percent solution. *PLoS Biology*, 8(9), e1000485.

WDPA, 2010. *The World Database on Protected Areas (WDPA)*, Gland & Cambridge: IUCN, UNEP-WCMC.

Western, D., Russell, S. & Cuthill, I., 2009. The status of wildlife in protected areas compared to non protected areas in Kenya. *PLoS ONE*, 4(7), e6140.

Whyte, I. J., 2005. *Census results for the elephant and buffalo in the Kruger National Park in 2005: Scientific Report 05/2005*, Skukuza: South African National Parks.

Woodley, S. et al., 2012. Meeting Aichi target 11: What does success look like for protected area systems. *PARKS*, 18(1), p. 23–36.

Wright, S. J., Sanchez-Azofeifa, G. A., Portillo-Quintero, C. & Davies, D., 2007. Poverty and corruption compromise tropical forest reserves. *Ecological Applications*, 17(5), p. 1259–1266.

WWF, 2012. *Living planet report 2012*, Gland: WWF.

WWF, 2014. *Living planet report: Species and spaces, people and places*, Gland: WWF.

Part III
Managing Protected Areas at System Scales

9

Toward Assessing the Vulnerability of US National Parks to Land Use and Climate Change

A. Hansen[1], *C. Davis*[2], *J. Haas*[3] *and N. Piekielek*[4]

[1]Ecology Department, Montana State University, Bozeman, MT, USA
[2]College of Forestry and Conservation, University of Montana, Missoula, MT, USA
[3]USDA Forest Service, Rocky Mountain Research Station, Fort Collins, CO, USA
[4]University Libraries, 208L Paterno Library, The Pennsylvania State University, University Park, PA, United States

Introduction

European colonization of the new world led to rapid conversion of wildlands to agricultural and urban landscapes. The desire to protect remaining natural lands, in the late 1800s, emerged as the modern concept of protected areas. The argument for formal landscape protection was that by removing the influence of humans, natural ecosystems would continue to maintain native species and natural ecological processes (Gaston et al., 2008). There is growing evidence, however, that protected areas are not presently functioning as originally envisioned. Ecological change within protected areas has been brought about by land use intensification on surrounding lands (DeFries et al., 2005; Wade & Theobald, 2009; Radeloff et al., 2010; Piekielek & Hansen, 2012). Introduction of invasive species has resulted in the extinction of native species within protected area boundaries (Newmark, 1985; Parks & Harcourt, 2002; Sanchez-Azofeifa et al., 2003). And in recent years, the effects of human-induced climate change within protected areas have become apparent on such variables as decreasing snowpack, increased rates of disturbance, prevalence of pests, and forest die-off (Loarie et al., 2009).

Protected Areas: Are They Safeguarding Biodiversity?, First Edition.
Edited by Lucas N. Joppa, Jonathan E. M. Baillie and John G. Robinson.

The extent and rate of such global changes differ across the Earth's surface as does the sensitivity of ecosystems to these changes (Running et al., 2004). Concomitantly, protected areas differ in their vulnerability to land use change, invasive species, and climate change. There is a need to assess vulnerability across networks of protected areas to determine those most at risk and to develop management and adaptation strategies that are tailored to local conditions and desired outcomes.

A promising framework for developing climate change adaptation strategies was recently developed by Glick et al. (2011). The four steps of the framework are:

1. Identify conservation targets.
2. Assess vulnerability.
3. Identify management options.
4. Implement management options.

The vulnerability assessment in step 2 includes four components (Turner II et al., 2003; IPCC Secretariat, 2007; Glick et al., 2011). "Exposure" is the degree of change in climate and land use, which are key drivers of ecological processes and biodiversity. "Sensitivity" is the degree to which species and ecological processes respond to a given level of exposure, largely based on the environmental tolerances of organisms. "Potential impact" results from the interaction of exposure and sensitivity. "Adaptive capacity" is the ability of a system to adjust to climate and non-climate change. It is the interaction of potential impact and adaptive capacity that determines "vulnerability."

Assessing vulnerability across networks of protected areas is a substantial challenge for several reasons. First, the managers of individual protected areas are typically so consumed with within-park issues such as managing visitor use that they often do not focus on the effects of land use changes outside their boundaries. Second, it is often unclear what parts of the surrounding landscape are strongly connected to the protected area and should be monitored and managed. Third, datasets for quantifying the four components of vulnerability are typically not available. Fourth, the spatial scales of analysis needed for assessing vulnerability to climate change are at a regional or subcontinental scale and not easily applied to conditions in a given ecosystem. Most protected area staff members are not well equipped to perform analyses at these scales. Fifth, protected areas are often considered as individual units and are managed in isolation from other protected areas.

Viewing individual protected areas as components of a network and assessing vulnerability across entire networks can reveal which protected areas face similar threats and allow synergies in adaptation strategies among them (Davis & Hansen, 2011). For example, those areas with low exposure and sensitivity may be best managed under the philosophy of maintaining "naturalness," more specifically, under the strategies of Natural Regulation (Boyce, 1998) or Historic Range of Variation (Keane et al., 2009). In contrast, protected areas with high exposure, high sensitivity, and low adaptive capacity are expected to undergo dramatic ecological change that will exceed any historic condition. Management in such cases should be aimed at promoting ecological integrity and resilience under changing future conditions (Hobbs et al., 2010). The goal of this chapter is to illustrate the initial steps in an assessment of vulnerability to land use and climate change for the network of national parks in the contiguous United States. The objectives are to:

- Delineate the lands surrounding protected areas where land use change can most impact protected areas.
- Quantify the exposure of protected areas to land use and climate change over the last century.
- Summarize potential protected area exposure and impact in the coming century.
- Consider implications for protected area management.

Delineating PACEs

Many protected areas were originally designated for reasons of scenic or recreational value, rather than any concern for ecological completeness or its adaptive capacity to respond to change. Thus, they may not include adequate area that is needed to maintain organism populations and essential ecological processes. Ecological processes such as nutrient flows, organism movements, disturbance regimes and population dynamics may operate over areas larger than the park itself. Land use intensification outside of protected areas may disrupt these flows and alter ecological processes and biodiversity within parks. Hansen & DeFries (2007) drew on ecological theory to identify mechanisms linking land use in surrounding lands with the ecological function of protected areas (Hansen & DeFries, 2007; Hansen et al., 2011). These mechanisms are summarized below:

- **Effective size:** Changing land uses may destroy natural habitats and reduce the effective size of the larger ecosystem, which can simplify the trophic structure as species with large home ranges are extirpated, because the size of the ecosystem to fall below that needed to maintain natural disturbance regimes, and reduce species richness through the loss of habitat area.
- **Ecological flows:** Changing land uses may alter characteristics of the atmosphere (climate, pollution), water (quantity, quality, nutrients, waterborne organisms), and natural disturbance (frequency, size, intensity) moving through the protected area.
- **Crucial habitats:** Changing land uses may eliminate or isolate crucial habitats, such as seasonal habitats, migration habitats, or habitats that support source populations.
- **Edge effects:** Changing land uses may increase human activity along park borders and result in the introduction of invasive species, increased hunting and poaching, and higher incidence of wildlife disturbance.

Hansen et al. (2011) measured these four mechanisms to quantitatively assess the impact of changing land uses on the ecosystems containing protected areas. They delineated protected area centered ecosystems (PACEs) around 13 US national park units using comprehensive scientific methods to map and analyze land use change within these PACEs. The PACEs were delineated based on five ecological criteria: Contiguity of surrounding natural habitat, watershed boundaries, extent of human edge effects, crucial habitats, and disturbance regimes (Davis & Hansen, 2011). The resulting PACEs were on average 6.7 times larger than the parks located in upper portions of watersheds and 44.6 times larger than those located in middle portions of watersheds. These results illustrated that human activities across relatively large areas surrounding protected areas may impact ecological factors in the protected areas. PACEs in the eastern United States were dominated by private lands with high rates of land development, suggesting that they offer the greatest challenge for management. Davis & Hansen (2011) used a similar method to map PACEs around the 57 largest US national parks in the contiguous United States and quantified land use change for the period 1940 to the present (Table 9.1). We draw on the results of these analyses in this chapter as a measure of exposure of 48 of these PACEs to land use change. Similarly, Haas (2010) quantified climate change during the past century within these PACEs, and we also summarize those results here (Haas, 2010).

Exposure to land use and climate change in the last century

Human activities have altered land use and climate in and around protected areas over the century since North American parks were established. This exposure to past changes may have pushed some protected areas closer to sensitivity thresholds and increased their vulnerability to recent changes. Moreover, spatial variation in past exposure among protected areas may be an indication of vulnerability under future land use and climate change. Thus, we summarize here past land use, invasion by nonnative species, and climate change for the US national parks and associated PACEs.

Davis & Hansen (2011) used six variables to determine the current level and extent of development: population density, housing density, and percentages of land with impervious surfaces, in agriculture, covered by roads, or in public land. Time series of population density, housing density, and land in agriculture were used to analyze changes over time. Cluster analysis was used to determine if patterns in major land use typologies could be distinguished. The authors found that population density within PACEs increased on average 224% from 1940 to 2000 and housing density increased by 329%, both considerably higher than average rates of change nationally. On average, 24% of private lands in PACEs were in exurban and rural housing densities. Individual PACEs differed substantially in area of public versus private (and thus developable) lands, the percent of the

Table 9.1 **Land use properties of the PACEs surrounding the US national park units included in this study.**

Park	PACE code	Date of park establishment	Park area (km^2)	PACE:park area ratio	PACE in public land (%)	Undeveloped private, 2000 (%)	PACE typology
Arches	ARCH	1929	309	29	93	88.9	Wildland protected
Badlands	BADL	1929	982	15	20	95.3	Wildland developable
Big Bend	BIBE	1935	3,291	8	20	99.6	Wildland developable
Bighorn Canyon	BICA	1964	484	33	33	96.1	Wildland developable
Big South Fork	BISO	1974	496	32	27	56.0	Exurban
Big Thicket	BITH	1974	359	15	5	63.4	Wildland developable
Blue Ridge Parkway	BLRI	1936	366	81	15	19.6	Exurban
Buffalo River	BUFF	1972	389	31	23	64.0	Agriculture
Canyon de Chelly	CACH	1931	375	18	6	92.3	Wildland developable
Colorado River*	CORI†	1908–1964	18,295	5	76	93.6	Wildland protected
Crater Lake	CRLA	1902	736	7	85	92.9	Wildland protected
Craters of the Moon	CRMO	1924	1,901	9	76	93.8	Wildland protected
Death Valley	DEVA	1933	13,764	4	79	80.7	Wildland protected
Delaware Water Gap	DEWA	1965	278	24	20	15.0	Exurban
Dinosaur	DINO	1915	853	22	74	94.7	Wildland protected
El Malpais	ELMA	1987	473	17	38	96.1	Wildland developable
Everglades, Big Cypress	EVER	1934, 1974	9,179	3	62	57.0	Urban
Glacier	GLAC	1910	4,080	5	80	77.9	Wildland protected
Great Basin	GRBA	1922	312	21	93	95.7	Wildland protected
Great Sand Dunes	GRSA	1932	496	18	43	91.1	Wildland developable
Great Smoky Mountains	GRSM	1926	2,098	7	43	15.1	Exurban
Guadalupe Mountains	GUMO	1966	356	21	44	99.5	Wildland developable
Joshua Tree	JOTR	1936	3,211	7	74	62.9	Wildland protected
Lake Roosevelt	LARO	1946	424	39	22	91.3	Agriculture
Lassen Volcanic	LAVO	1907	434	9	47	92.2	Wildland developable
Missouri River	MNRR	1978	279	71	6	82.4	Agriculture
Mojave	MOJA	1994	6,433	3	94	88.9	Wildland protected

(*Continued*)

Table 9.1 **(Continued)**

Park	**PACE code**	**Date of park establishment**	**Park area (km^2)**	**PACE:park area ratio**	**PACE in public land (%)**	**Undeveloped private, 2000 (%)**	**PACE typology**
Mount Rainier	MORA	1899	952	6	70	80.2	Wildland protected
New River Gorge	NERI	1978	285	28	15	42.1	Exurban
North Cascades Complex	NOCA	1968	2,756	6	91	73.8	Wildland protected
Olympic	OLYM	1909	3,700	5	47	65.1	Wildland developable
Organ Pipe Cactus	ORPI	1937	1,338	8	43	99.1	Wildland developable
Ozark	OZAR	1964	333	29	29	82.8	Wildland developable
Petrified Forest	PEFO	1906	903	11	25	97.3	Wildland developable
Pictured Rocks	PIRO	1966	298	17	60	82.5	Wildland developable
Point Reyes, Golden Gate	POGO	1962, 1972	617	10	28	32.6	Urban
Redwood	REDW	1968	468	15	60	82.2	Wildland developable
Rocky Mountain	ROMO	1915	1,080	8	83	53.6	Wildland protected
Saint Croix	SACN	1968	396	21	15	66.5	Agriculture
Saguaro	SAGU	1933	378	48	55	57.2	Exurban
Santa Monica Mountains	SAMO	1978	619	9	33	27.6	Urban
Shenandoah	SHEN	1926	782	14	16	19.8	Exurban
Sleeping Bear Dunes	SLBE	1970	284	16	36	22.7	Exurban
Theodore Roosevelt	THRO	1947	285	30	41	95.8	Agriculture
Voyageurs	VOYA	1971	829	11	77	80.6	Wildland protected
White Sands	WHSA	1933	617	17	34	74.3	Wildland developable
Yellowstone Grand Teton	YELL	1872	10,159	3	93	73.3	Wildland protected
Yosemite, Sequoia-Kings Canyon	YOSE	1890	6,521	3	90	66.2	Wildland protected
Zion	ZION	1909	598	14	69	87.1	Wildland protected
Mean			2,140	18	49	72.62	

Source: From Hansen et al. (2014).

*Colorado River parks are: Canyonlands, Capitol Reef, Glen Canyon, Grand Canyon, and Lake Mead.

private lands that had been developed, and the land use type and housing densities on those private lands.

Five categories of PACEs with similar patterns of development were distinguished using statistical clustering analysis (Figure 9.1):

- Wildland protected
- Wildland developable
- Agriculture
- Exurban
- Urban

PACEs within a category tended to occur in particular regions of the country. For example, both types of wildland PACEs were in the mountain west and southwest deserts. Agriculture and exurban PACEs were in the Great Plains and eastern United States. The unique management challenges and opportunities faced by each group of PACEs were identified, with the goal of allowing managers to collaborate with each other based on similarity of challenges that they face. This is the

Figure 9.1 **Protected area centered ecosystems (PACEs) surrounding each US national park, color-coded by land use typological membership. Classification criteria were as follows: wildland protected, >65% public; wildland developable, <65% public, >60% private undeveloped, <16% private agriculture; agricultural, <65% public private, >60% undeveloped, >16% private agriculture; exurban, <65% public private, >60% undeveloped, <15% private dominated by exurban or urban; urban, <65% public private, >60% undeveloped, >15% private dominated by urban or urban. Source: From Davis & Hansen (2011). © Ecological Society of America**

first effort to develop a typology of protected areas based on land use change in the surrounding ecosystem. Other networks of protected areas may find this methodology useful for prioritizing monitoring, research, and management among groups with similar vulnerabilities and conservation issues.

Haas (2010) quantified trends in climate across these PACEs for the period 1985–2009. She drew on the precipitation elevation regressions on independent slope model (PRISM) climate dataset (Daly et al., 2002), which uses observations from meteorological stations to produce a spatially continuous 4 km surface of spatially interpolated climate values. These data were used to estimate 100-year trends in mean annual temperature, mean annual precipitation, and a moisture index derived from precipitation and potential evapotranspiration (a function of temperature and solar radiation). Results indicated that 78% of the PACEs warmed significantly (~1°C per 100 years). The highest rates of warming (2°C per 100 years) occurred in the mountain west. Precipitation increased significantly in 22% of the PACEs, largely in the midwest United States. The moisture index revealed that 17% of the PACEs increased in water balance while 3% decreased in water balance, largely due to increased evapotranspiration due to warming.

There were also biological changes associated with the changes in land use and climate over time. Invasive species are a good indicator of stress in protected area ecosystems. Hansen et al. (2014) found that the proportion of vascular plants that were nonnative correlated with the land use change typology. Wildland PACEs (protected and developable) had the lowest proportion of nonnatives, agricultural and exurban PACEs had intermediate levels, and urban PACEs had the highest proportions.

The relative magnitude of past change in climate, land use, and invasive species for each PACE was calculated as the percentage of the highest value among the PACEs for each variable. These percentages were summed to represent the relative magnitude of the combined exposure to these three components of global change (Figure 9.2). Notice that some PACEs had high rates of warming; others had high

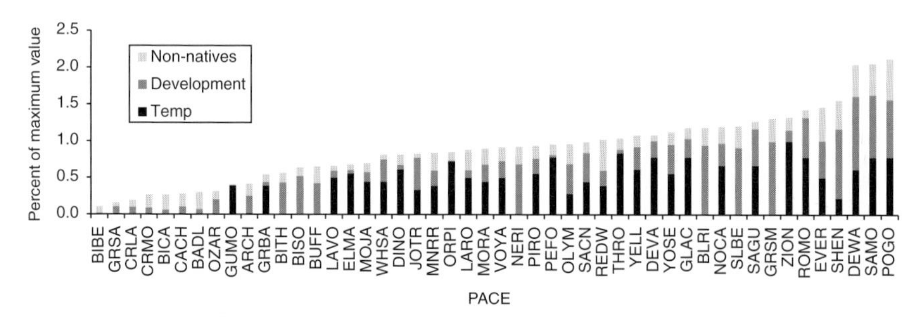

Figure 9.2 **Relative levels of temperature change, proportion of PACE developed, and nonnative vascular plants among PACEs. The data are expressed as the percentage of the highest value among the PACEs for each variable, and these percentages are summed to represent the relative magnitude of the combined exposure to these three components of global change**

rates of development and invasive species, while some had high rates of all three, and a few PACEs had relatively little change in any of the three factors. This illustrates how exposure can differ dramatically among PACEs. PACEs such as Santa Monica and Point Reyes in southern California and Delaware Water Gap and Everglades in the eastern United States had total rates of change more than five times those in Great Sand Dunes and several other western PACEs.

Forecasting change in climate and biome location: 2010–2090

Climate change over the past century has altered or disrupted ecosystems in some geographic locations. In the western United States, increased temperature and drought stress have led to reduced snowpack and runoff, increasingly severe fire and pest outbreaks, and forest die-off (US Global Change Research Program, 2009). Spatial variation in ecosystem response to climate change is a result of variation in rates of climate change and in the key factors that limit plant growth. Across terrestrial biomes, plant growth is generally limited by low temperatures across northern latitudes, water shortage in middle latitudes, and radiant energy in the humid tropics (Running et al., 2004). Individual plant species exhibit specific tolerances to climate conditions. It is when climate change exceeds the tolerances of plant species that widespread plant mortality can occur (Allen et al., 2010). In many places around the earth that are water limited, increased drought has resulted in forest die-off (Anderegg et al., 2012). Future climate change and vegetation mortality are expected to lead to major shifts in the location of biome types. Such biome shifts are particularly worrisome with regard to protected areas because climate conditions within them may become unsuitable for ecosystems that define them such as the potential for the giant sequoia to no longer inhabit Sequoia National Park. In this section, we report on the projected change in temperature and biome type across US national parks.

Rehfeldt et al. (2012) used current climate and downscaled global climate model projections to develop climate niche models for biomes of North America and to project change in area and distribution under six climate change scenarios (Rehfeldt et al., 2012). Projected future temperature under a midrange emissions scenario indicated that individual PACEs will warm between 1.1 and 2.5°C by 2030, a magnitude slightly in excess of the average 1°C rise reported by Haas (2010) over the past century and a rate more than three times faster than has been experienced in the recent past (Figure 9.3). By 2090, mean annual temperatures are projected to be 3.4–6.0°C warmer than present. PACEs with the highest projected warming are in the upper midwestern United States and in the western mountains and deserts. Projected temperature increases in these locations are up to twice as great as in PACEs on the east and west coasts of the United States, where maritime effects are thought to moderate climate change.

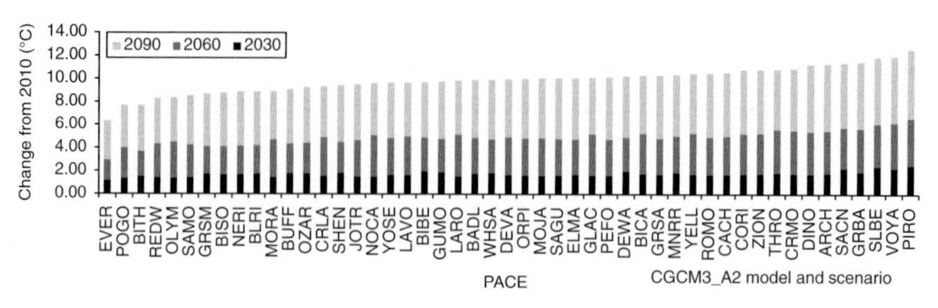

Figure 9.3 **Projected change in mean annual temperature within protected area centered ecosystems (PACEs) surrounding US national parks under the CGCM3_A2 model and scenario. Source: Data from Rehfeldt et al. (2012)**

In order to gauge the possible response of vegetation to projected climate change in these PACEs, Hansen et al. (2014) used Rehfeldt's (2012) maps of projected change in biome suitability to quantify the proportion of each PACE that is projected to undergo a shift. The results strikingly illustrate how the potential impact of climate change may vary among protected areas. Many PACEs in the midwestern and eastern United States are projected to undergo less change in biome type by 2090 than PACEs in the mountain west by 2030 (Figure 9.4). Note that the four PACEs in the upper midwest that have high rates of projected future warming undergo little projected change in biome type. Vegetations in these areas are more temperature limited than water limited and are not expected to undergo drought stress under future climates. It is drought stress that drives the projected strong biome shifts in many western mountain and desert PACEs such as Zion.

More generally, ecosystems with the higher sensitivity to climate change are expected to be those supporting species with narrow climate tolerances, locations near the edge of zones of climate suitability, and species and communities subject to higher disturbance rates (e.g., fire, pests). Ecosystems also differ in adaptive capacity. Those dominated by plants with shallow rooting, short life spans, and limited dispersal abilities are expected to be less able to adapt to changing climate conditions. Alternatively, forests that are long-lived, deeply rooted, and resilient to disturbance may be able to persist for lengthy periods of time at a site that has undergone substantial climate change even when future climate is no longer suitable.

Cumulative effects of past and projected future climate and land use

Land use change is expected to reduce the ability of ecosystems to adjust to changing climate. Land use is associated with increased invasion by exotics, for example, which may take over sites under changing climates. Land use also reduces

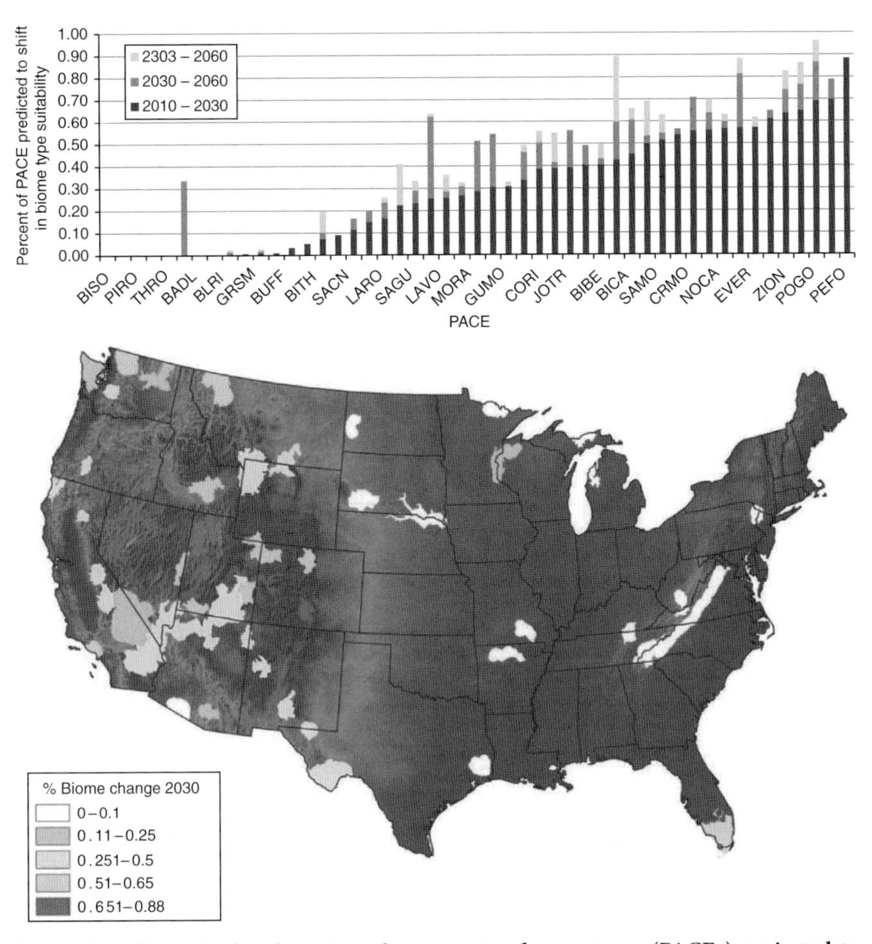

Figure 9.4 **Percent of each protected area centered ecosystems (PACEs) projected to shift in biome type suitability under the consensus of six climate models and scenarios (above) and color-coded level of biome suitability shift by 2030 mapped across the United States. Source: From Hansen et al. (2014)**

connectivity across the landscape and thus reduces the ability of native species to disperse to newly suitable habitats. Thus, the potential vulnerability of PACEs needs to be assessed by considering both past land use and potential impact of future climate. Hansen et al. (2014) positioned each PACE in the space defined by percentage of the PACE developed in 2010 and by the percentage of the PACE projected to undergo a biome shift by 2030 (Figure 9.5). This plot reveals that many PACEs are low in both land use intensity and potential climate impact, others are high in either land use intensity or potential climate impact, and a few PACEs are high in both land use intensity and potential climate impact.

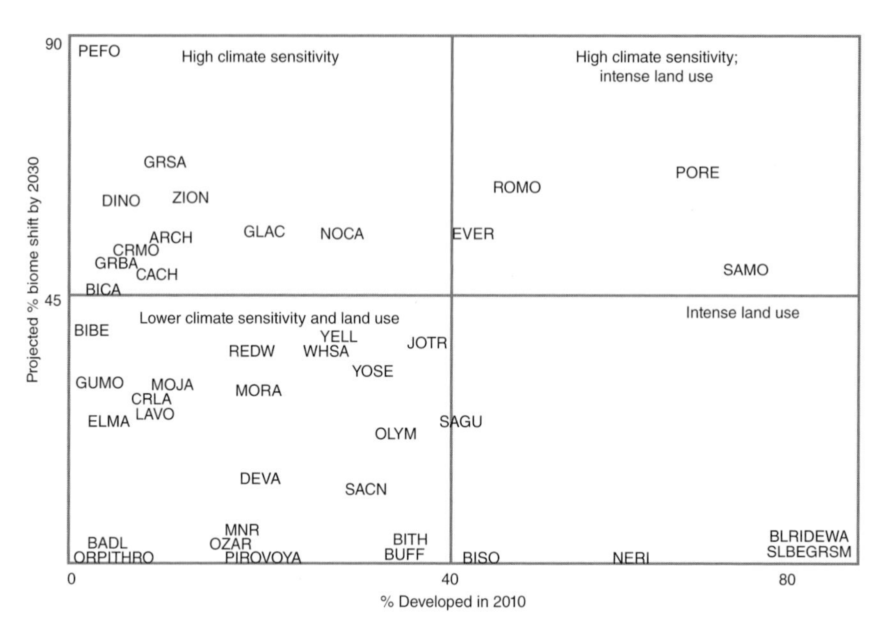

Figure 9.5 **PACEs positioned in the space defined by the percentage of the PACE developed in 2010 and by the percentage of the PACE projected to undergo a biome shift by 2030. Source: From Hansen et al. (2014)**

Implications for management

The third step in climate adaptation planning (Glick et al., 2011) is to evaluate management options. Knowledge of the components of vulnerability provides a basis for evaluating the need for management and the management approaches that might be most effective. Places that are low in exposure to climate and land use change, low in sensitivity, high in adaptive capacity, and thus low in vulnerability may be a relatively low priority for global change adaptation management (Hansen et al., 2014). Current ecological conditions are likely to persist in such places. Alternatively, those that are highly vulnerable may be difficult to save through management actions. Such ecosystems are likely to cross ecological thresholds and undergo changes of state in ecological function and biodiversity. Management of such ecosystem may be best aimed at encouraging adaptation to new conditions. Ecosystems of moderate vulnerability may be the highest priority for active management strategies to maintain current conditions.

What management philosophies can best help us achieve ecological objectives under climate and land use change? There is currently a vigorous debate between advocates for managing for "naturalness" versus advocates for managing for change to impending future conditions. Keane et al. (2009) represent the "naturalness" position. Termed the Historic Range of Variation approach or the Ecological Process

Management, the goal is to perpetuate natural pattern and ecological function in the face of change. The general management strategy for achieving this goal is to maintain landscape patterns and ecological processes within the range of variation in place prior to European influence. Under this philosophy, active management is generally dissuaded for fear of causing harm to the ecosystem or compromising our ability to learn from the effects of change in unmanaged ecosystems (i.e., parks as natural laboratories). Active management is used only where needed to restore lost ecological patterns or functions. Reintroducing an extirpated keystone predator would be an example of this management philosophy as was done with the reintroduction of the gray wolf into Yellowstone National Park (Boyce, 1998).

In contrast to the "naturalness" point of view, Hobbs et al. (2010) argue vigorously that many ecosystems have been or will be so altered by land use and climate change that current patterns and function will have little resemblance to historical conditions (Hobbs et al., 2010). Consequently, the goal of management should be to promote ecological integrity and resilience under future changing conditions. This "first principles" approach advocates designing future ecosystems based on guiding principles, including historical fidelity, autonomy of nature, ecological integrity and resilience, as well as managing with humility. Active management strategies such as translocating species to newly suitable habitats to encourage range shifts would be an example of implementing this approach.

Hansen et al. (2014) suggested that the vulnerability of protected areas to global change can be used as a guide to the management philosophy that is most appropriate for a given area. "Historic range of variation" or "natural process management" would seem appropriate for protected areas with low vulnerability to global change. Patterns and dynamics typical of pre-European settlement are expected to continue into the future in such systems. Management to keep the system within this historic range is feasible and desirable. An example may be Olympic National Park on the Pacific coast of the northwest United States. Past climate and land use change in this PACE have been low to moderate. Future climate change might be low relative to other PACEs due to the moderating influence of the Pacific Ocean. The tall and long-lived rain forest species that dominate this PACE are resilient to shifts in climate and disturbance. Managers in this region are more concerned about maintaining disturbance regimes and the full suite of several stages required for high biodiversity in this ecosystem than they are about the effects of climate change. Historic range of variation continues to be the guiding philosophy of management in this region.

The Santa Monica PACE near Los Angeles in southern California, in contrast, shows high vulnerability to global change and management for future conditions may be the only viable approach. This national park is surrounded by urban areas, with some 72% of the PACE being developed. Nearly a quarter of the vascular plants are nonnative. Temperature has warmed substantially in the past century and this trend is projected to continue in the coming century. More than 50% of the PACE is projected to undergo a biome shift by 2030. Clearly, management to retain pre-European disturbance (such as natural fire regimes) or ecological conditions in this ecosystem would be either futile or socially unacceptable. Instead, managers here must decide what ecological conditions and ecosystem services they wish to achieve under future global change and develop active management strategies to produce and maintain this condition and these services.

Most PACEs in the United States lie somewhere between these two extremes in vulnerability. For such PACEs, some combination of the two management philosophies may be most appropriate. Both Keane et al. (2009) and Hobbs et al. (2010) conclude that both looking at past conditions and considering future conditions can inform management.

Concluding remarks

Maintaining native biodiversity and ecological function in protected areas is an enormous challenge in a world increasingly dominated by humans. Our key conclusion is that protected areas potentially differ in each of the components of vulnerability to land use and climate change and that knowledge of these differences can be used to guide the development of locally appropriate adaptation strategies. A first step in quantifying vulnerability of protected areas is to delineate the surrounding lands that are strongly connected to the ecological functioning of the protected area and summarizing past and projected future land use and climate change in these PACEs. Within 48 PACEs focused on US national parks, we found that exposure to land use and climate change and sensitivity of vegetation and potential impact on biome types varied dramatically. Among these PACEs, some showed high components of vulnerability to land use or potential impacts on biome shifts due to climate change or both. While our analysis to date did not deal with all components of past and future vulnerability, it illustrates the utility of such approaches for designing adaptation strategies. PACEs with low vulnerability may best be managed with the well-developed strategies related to maintaining "naturalness." PACEs with high vulnerability are candidates for strategies that focus on maintaining ecological function under the future conditions that the PACE is expected to face. This approach to assessing vulnerability should be applied to other networks of protected areas around the world to provide a basis for effective management and adaptation.

Acknowledgments

Funding was provided by the NASA Applied Sciences Program and the NASA Land-Cover/Land-Use Change Program. Data were provided by Gerald Rehfeldt, USDA Forest Sciences Lab, Moscow, ID.

References

Allen, C. D. et al., 2010. A global overview of drought and heat-induced tree mortality reveals emerging climate change risks for forests. *Forest Ecology and Management*, Volume 259(4), p. 600–684.

Anderegg, W. R. L. et al., 2012. The roles of hydraulic and carbon stress in a widespread climate-induced forest die-off. *Proceedings of the National Academy of Sciences of the United States of America*, Volume 109, p. 233–237.

Boyce, M. S., 1998. Ecological process management and ungulates: Yellowstone's conservation paradigm. *Wildlife Society Bulletin*, Volume 26(3), p. 391–398.

Daly, C. et al., 2002. A knowledge-based approach to the statistical mapping of climate. *Climate Research*, Volume 22, p. 99–113.

Davis, C. R. & Hansen, A. J., 2011. Trajectories in land use change around US national parks and their challenges and opportunities for management. *Ecological Applications*, Volume 21(8), p. 3299–3316.

DeFries, R., Hansen, A. J., Newton, A., & Hansen, M. C., 2005. Isolation of protected areas in tropical forests over the last twenty years. *Ecological Applications*, Volume 15(1), p. 19–26.

Gaston, K. J., Jackson, S. F., Cantu-Salazar, L., & Cruz-Pinon, G., 2008. The ecological performance of protected areas. *Annual Review of Ecology, Evolution and Systematics*, Volume 39, p. 99–113.

Glick, P., Stein, B. A., & Edelson, N. A., 2011. *Scanning the conservation horizon: A guide to climate change vulnerability assessment*, Washington, DC: National Wildlife Federation.

Haas, J., 2010. Quantifying trends in our national parks: A landscape level analysis of climate change and ecosystem productivity. MSc Thesis. Bozeman: Montana State University.

Hansen, A. J., & DeFries, R., 2007. Ecological mechanisms linking protected areas to surrounding lands. *Ecological Applications*, Volume 17(4), p. 974–988.

Hansen, A. J. et al., 2011. Delineating the ecosystems containing protected areas for monitoring and management. *BioScience*, Volume 61(5), p. 363–373.

Hansen, A. J. et al., 2014. Exposure of US national parks to land use and climate change 1900–2100. *Ecological Applications*, Volume 24, 484–502.

Hobbs, R. J. et al., 2010. Guiding concepts for park and wilderness stewardship in an era of global environmental change. *Frontiers in Ecology and the Environment*, Volume 8, p. 483–490.

IPCC Secretariat, 2007. Climate Change 2007: Impacts, adaption and vulnerability. Working Group II—Contribution to the Intergovernmental Panel on Climate Change, Fourth Assessment Report. Geneva: IPPC Secretariat.

Keane, R. E., Hessburg, P. F., Landres, P. B., & Swanson, F. J., 2009. The use of historical range and variability (HRV) in landscape management. *Forest Ecology and Management*, Volume 258, p. 1025–1037.

Loarie, S. R. et al., 2009. The velocity of climate change. *Nature*, Volume 462, p. 1052–1055.

Newmark, W. D., 1985. Legal and biotic boundaries of western North American national parks: A problem of congruence. *Biological Conservation*, Volume 33, p. 197–208.

Parks, S. A. & Harcourt, A. H., 2002. Reserve size, local human density, and mammalian extinctions in the US protected areas. *Conservation Biology*, Volume 16, p. 800–808.

Piekielek, N. B. & Hansen, A. J., 2012. Extent of fragmentation of coarse scale habitats in and round US national parks. *Biological Conservation*, Volume 155, p. 13–22.

Radeloff, V. C. et al., 2010. Housing growth in and near United States protected areas limits their conservation value. *Proceedings of the National Academy of Sciences of the United States of America*, Volume 107(2), p. 940–945.

Rehfeldt, G. E., Crookston, N. L., Saenz-Romero, C., & Campbell, E. M., 2012. North American vegetation model for land use planning in a changing climate: A solution to large classification problems. *Ecological Applications*, Volume 22(1), p. 119–141.

Running, S. W. et al., 2004. A continuous satellite derived measure of global terrestrial primary production. *BioScience*, Volume 54, p. 547–560.

Sanchez-Azofeifa, G., Daily, G. C., Pfaff, A. S., & Busch, C., 2003. Integrity and isolation of Costa Rica's national parks and biological reserves: Examining the dynamics of land cover change. *Biological Conservation*, Volume 109, p. 123–135.

Turner II, B. L. et al., 2003. A framework for vulnerable analysis in sustainability science. *Proceedings of the National Academy of Sciences of the United States of America*, Volume 100, p. 8074–8079.

US Global Change Research Program, 2009. Global Climate Change Impacts in the United States: 2009 Report. [Online] Available at: http://nca2009.globalchange.gov/ (accessed September 7, 2015).

Wade, A. A. & Theobald, D. M., 2009. Residential development encroachment on US protected areas. *Conservation Biology*, Volume 24(1), p. 151–161.

10

Integrating Community-Managed Areas into Protected Area Systems: The Promise of Synergies and the Reality of Trade-Offs

M. Rao[1], *H. Nagendra*[2], *G. Shahabuddin*[3] *and L. R. Carrasco*[4]

[1]Wildlife Conservation Society, Bronx, NY, USA
[2]School of Development, Azim Premji University, Bangalore, India
[3]Centre for Ecology, Development and Research, Uttarakhand, India
[4]Department of Biological Sciences, National University of Singapore, Singapore

Introduction

The Convention on Biological Diversity (CBD), the most relevant international policy instrument addressing protected areas, recognizes their significance for the conservation of biodiversity. At the 10th meeting of the Conference of the Parties to the CBD in Nagoya, Japan, governments committed to expand the protected area coverage to 17% of the earth's surface. At the same meeting, governments adopted the revised Strategic Plan for Biodiversity (2011–2020) including the 20 Aichi Biodiversity Targets (CBD, 2010).

While extent of protected areas has increased over the past few decades, biodiversity continues to be lost. A recent report by Butchart et al. (2010) showed indicators of biodiversity status trending negatively (as measured by species' population trends, extinction risk, and habitat extent), which raises questions about the effectiveness of protected areas to safeguard biological diversity. Many factors (e.g., design, management, and financing) affect the effectiveness of protected areas, but the issue of governance is of predominant importance (Butchart et al., 2010).

Protected area governance varies: there are protected areas strictly protected by the state where people are excluded to areas intensively used and managed by local communities, protected areas where the state has complete decision-making authority, and protected areas with comanagement and shared decision making by

Protected Areas: Are They Safeguarding Biodiversity?, First Edition.
Edited by Lucas N. Joppa, Jonathan E. M. Baillie and John G. Robinson.

government agencies and other stakeholders (Bertzky et al., 2012). In this chapter, we are concerned with indigenous and community conserved areas (CCAs), defined as natural or modified ecosystems containing biodiversity values, ecological services, and cultural values. They are managed by indigenous and other communities through local or customary laws and are found in both terrestrial and marine areas, ranging in size from less than 10,000 m^2 sacred groves to greater than 30,000 km^2 indigenous territories in Brazil (Oviedo, 2006; Berkes, 2009). By 2010, the World Database on Protected Areas (WDPA) recorded some 700 protected areas known to be governed by indigenous peoples and or local communities, covering over 1.1 million km^2 or 9.3% of the total protected area with a known governance type (Bertzky et al., 2012).

There has been growing emphasis on the need to recognize the relevance of biologically important areas managed by indigenous and local communities to national protected area systems (Borrini-Feyerabend et al., 2004). CCAs have been viewed as mechanisms to strengthen the management and expand the coverage of the world's protected areas, to promote connectivity, and to enhance public support for protected areas. CCAs constitute an integral part of this plan and are relevant to achieving key global biodiversity targets. For example, Aichi Target 11 aims for "equitably managed" systems of protected areas and "other effective area-based conservation measures," including indigenous peoples' territories and CCAs. Further, Aichi Target 18 aims for the incorporation of "traditional knowledge, innovations, and practices of indigenous and local communities" in the Convention through "full and effective participation." Given that communities govern significantly large areas, it becomes relevant to more closely examine the contribution of CCAs in conserving biological diversity and in helping achieve key Aichi Biodiversity Targets.

CCAs as important solutions to biodiversity conservation: The assumptions

CCAs involve communities linked culturally and economically to the ecosystems in which they live (Berkes, 2009). Typically, the community is the major decision maker, and community institutions have the authority to enforce regulations (Pathak et al., 2004). While there is considerable variation across sites, the common principle of CCAs, in comparison with strictly protected areas (SPAs), is the emphasis on local management, access, and control over ecosystem goods and services. Increased global support for the community-managed paradigm stems in part from the perception of social injustices caused by the creation of SPAs that restrict local residence, access, and use (Brockington, 2003; Lewis, 2003).

The efficacy of CCAs to protect biodiversity depends on a number of key assumptions. First, when people are conferred ownership or management rights over biodiversity and ecosystem services from which they can legitimately benefit, they will support conservation objectives (Rai & Uhl, 2004). The second assumption is that communities often have deep knowledge of local ecology; consequently, their use of traditional harvesting techniques will tend to minimize ecological damage (Pathak et al., 2004). Traditional knowledge of low-damage extractive practices is thought to go along with a deeper conservation ethic that is often intertwined with religious taboos and restrictions (Byers et al., 2001). The third assumption is that management decisions of the community will effectively lead to conservation, even though conservation may not be the primary objective.

While CCAs are considered as an important solution to the problem of protecting natural diversity, the evidence for their effectiveness in conserving biological diversity remains scarcely examined (Berkes, 2007, 2009). This is problematic within the context of globally declining biodiversity and failure of nations to achieve biodiversity goals (Butchart et al., 2012). The biological effectiveness of different governance regimes needs to be assessed. In particular, given the limited coverage of state-managed protected areas, it becomes pertinent to understand what role CCAs play in achieving biodiversity targets, and this will be central to understanding their role in the design and implementation of global protected area systems.

Protected areas as complex socioecological systems: The relevance of institutional theory

Underlying community use and management of biodiversity within protected areas are complex issues of institutions and governance of common pool resources (Ostrom, 1990). Extensive research has yielded a body of theory describing the range of socioeconomic and environmental factors that affect the development, maintenance, and performance of common property institutions (Agrawal & Gibson, 1999; Berkes, 2010), such as typically managed CCAs. Ostrom (2007) identified all natural resources used by people as being embedded in complex, socioecological systems (SESs) composed of subsystems and internal variables within these subsystems at multiple levels (Ostrom, 2007). Other scholars, for example, Berkes (2007), noted that the panacea of community-based conservation could be expected to be no more effective than the panacea of exclusively state-based conservation because they both ignored the complex nature of horizontal and vertical linkages across organizations at local, regional, and international levels that underlie the conservation of biodiversity in protected areas.

SESs are complex, nested systems that operate at and need to be analyzed at multiple spatial, temporal, and organizational scales. Ostrom (2007) introduced a general SES framework to identify variables that affect the likelihood of self-organization in efforts to achieve a sustainable SES. This framework is aimed at enabling a diagnostic exercise to identify the most important challenges for successful governance of ecological resources. We apply this framework to examine key drivers of biological outcomes in CCAs.

The aim of this chapter is to examine the biological effectiveness of community governance regimes with the objective of examining broad patterns, strong determinants, and expected trade-offs. First, we provide a summary of comparative biological effectiveness of community-managed landscapes relative to landscapes managed under strict protection regimes based on a comprehensive literature review. Second, we apply Ostrom's (2007) SES framework to analyze key socioeconomic, environmental, and institutional drivers of biological outcomes in CCAs. Third, we provide a set of guiding principles for the integration of CCAs into national protected area systems.

A comparative analysis of biological outcomes in community-managed landscapes

The conservation effectiveness of community-based conservation can be tested by a quantitative comparison of biological indicators in CCAs versus those in SPAs and open-access (unmanaged) areas within the same ecoregion. Analyzing trends in the status of these indicators over time provides a further assessment. To examine this question, Shahabuddin & Rao (2010) identified a set of key biological indicators (Table 10.1) from the published literature and noted whether they were higher, lower, or similar to comparables (Shahabuddin & Rao, 2010).

Comparisons of community-managed/used areas with SPAs

In comparisons of community-managed areas with SPAs, forest cover, deforestation rates, species diversity/richness, and forest structure showed no clear differences between governance systems. CCAs were worse than SPAs for species composition and species abundance indicators. There is evidence of increased deforestation and loss of vegetative cover in areas managed or used by communities in Mexico (Johnson & Nelson, 2004), the Colombian Guyana Shield (Armenteras et al., 2009), and Tanzania (Pelkey et al., 2000). However, similar deforestation rates were recorded in inhabited protected areas and recently inhabited community forests in Guatemala and Mexico (Bray et al., 2008). In Thailand however, community-managed mangrove forests were in significantly better condition than state forests (Sudtongkong & Webb, 2008).

In contrast, in all seven ecosystems where species composition was studied, CCAs and SPAs were found to host distinct ecological communities. Gardner et al. (2007) examined several vertebrate groups located along a four-step gradient of increasing human activity from national park to a forest–village mosaic in Tanzania. While species richness did not vary much along this human-use gradient, suites of species within each major class of vertebrates were significantly

Table 10.1 **A summary of positive, negative, and neutral outcomes in biodiversity conservation from selected case studies**

Variable measured	Positive biological outcomes	Negative biological outcomes	Neutral outcomes
Comparison: CCAs with SPAs			
Forest cover	Nagendra (2007)	Pelkey et al. (2000)	Bray et al. (2008)
Deforestation rate	Nepstad et al. (2006) Ruiz-Perez et al. (2005)	Armenteras et al. (2009) Johnson & Nelson (2004)	Bray et al. (2008)
Species richness	Shackleton (2000) Garcia & Pascal (2005) Ranganathan et al. (2008)	Bossart et al. (2006) Wallgren et al. (2009) Fabricius et al. (2003)	Gardner et al. (2007)
Species diversity	Shackleton (2000)	Bossart et al. (2006) Wallgren et al. (2009) Dorji et al. (2003)	Sudtongkong & Webb (2008) Fabricius et al. (2003) Webb & Gautam (2001)
Basal area	Sudtongkong & Webb (2008)	Garcia & Pascal (2005) Webb & Gautam (2001)	
Forest height	Sudtongkong & Webb (2008)		
Stem/tree density		Sudtongkong & Webb (2008) Garcia & Pascal (2005)	
Tree regeneration			Johnson & Nelson (2004)
Canopy cover (%)		Johnson & Nelson (2004)	
Plant species abundance		Garcia & Pascal (2005)	Bhagwat et al. (2005)
Animal species abundance		Setsaas et al. (2007) Carillo et al. (2000) Ranganathan et al. (2008) Wallgren et al. (2009) Bhagwat et al. (2005)	

(*Continued*)

Table 10.1 **(Continued)**

Variable measured	Positive biological outcomes	Negative biological outcomes	Neutral outcomes
Forest structure		Webb & Gautam (2001)	
Threatened species	Bhagwat et al. (2005)	Wallgren et al. (2009)	
Comparison: CCAs with open-access areas			
Forest cover	Campbell (2004)	Pelkey et al. (2000)	Somanathan et al. (2009)
	Bajracharya et al. (2005)		Tengo et al. (2007)
Species diversity	Bajracharya et al. (2005)		
	Campbell (2004)		
Basal area	Bajracharya et al. (2005)	Sarkar (2008)	
	Campbell (2004)		
Forest height	Rai & Uhl (2004)	Sarkar (2008)	
Tree regeneration		Sarkar (2008)	
Canopy cover (%)		Campbell (2004)	
Grassland productivity	Leisher et al. (2012)		
Comparison: CCAs over time			
Forest cover		Tucker (2004)	Dalle et al. (2006)
		Pelkey et al. (2000)	
		Bray et al. (2008)	
		Peralta & Mather (2000)	
		Armenteras et al. (2009)	
		Pedlowski et al. (2005)	

Deforestation rate		Peralta & Mather (2000)	Nepstad et al. (2006)
		Perez-Verdin et al. (2009)	
		Byers et al. (2001)	
Basal area		Holder (2004)	
Forest height		Holder (2004)	
Tree density		Holder (2004)	
Sapling survival		Saha & Howe (2003)	
		Saha & Howe (2006)	
Plant species abundance	Wallgren et al. (2009)		
Animal species abundance	Nawaz (2008)	Carillo et al. (2000)	
		Peres & Nascimento (2006)	
		Franzen (2006)	

Source: Modified from Shahabuddin & Rao (2010). Reproduced with permission of Elsevier.

altered. Strong negative trends also held true of plant and animal indicator species of conservation importance, which were lower in CCAs (Carillo et al., 2000; Setsaas et al., 2007). In India, a network of sacred groves did not offer comprehensive protection to endemic species in comparison with an adjacent strictly protected wildlife sanctuary (Garcia & Pascal, 2005). In the Western Ghats, the abundance of evergreen tree species and forest-preferring bird species were most abundant in strictly protected landscapes followed by sacred groves and coffee plantations, in that order (Bhagwat et al., 2005; Ranganathan et al., 2008).

Comparisons of community-managed/used areas with open-access systems

In comparisons with open-access systems, CCAs play a more positive role in conserving species diversity (Bajracharya et al., 2005), richness (Campbell, 2004), and reducing deforestation. Open-access areas tend to be overexploited since no single user has an overriding stake in their conservation (Rai & Uhl, 2004). Sacred forests showed reduced forest loss and higher plant species richness in comparison to open-access systems (Byers et al., 2001; Campbell, 2004). In Madagascar, Tengo et al. (2007) showed that sacred forests protected endemic species not found in unprotected forests. However, quantitative comparisons of basal area, canopy cover, and regeneration status showed that Van Panchayats[1] forests in India have undergone considerable degradation during recent years, faring no better than government managed reserved forests that represent an open-access resource for all practical purposes (Sarkar, 2008; Somanathan et al., 2009).

Community-managed forests over time

Biological indicators tend to decline over time in community-managed forests. For example, forest management strategies increased the densities of the commercially valuable açaí palm, changing the structure and composition of varzea forests in the Amazon over time (Weinstein & Moegenburg, 2004). In Guatemala, unsustainable extraction of fuelwood resulted in decreased stand density and reduced basal area over a 40-year period in a community forest with customary rules for forest extraction (Holder, 2004). Fires used by local communities as a management tool to enable nontimber forest product collection negatively impacted the regeneration of several species of native trees (Saha & Howe, 2003, 2006). One of the few exceptions to the trend of declining biodiversity over time in CCA forests is the study completed by Dalle et al. (2006), which found negligible forest loss before and after the initiation of a community forestry program in the "ejidos" of

Mexico (Dalle et al., 2006). Over time, expansion of agricultural grazing lands and unsustainable levels of hunting and resource extraction seem to underlie habitat degradation in community-managed areas (Franzen, 2006).

That effectiveness of community-managed areas varied so greatly highlights the need to examine socioeconomic, environmental, and institutional factors that might influence the observed patterns in biological outcomes.

Biological outcomes in CCAs: Analysis of socioeconomic, environmental, and institutional drivers

We examined the drivers underlying the performance of community-managed areas in order to discern conditions associated with positive and negative outcomes for biodiversity in these areas. Fifty-six case studies (22 positive and 34 negative biological outcomes) were included in an analysis that considered key institutional, socioeconomic, and governance groups of variables defined within the SES framework (Ostrom, 2007). Variables included:

- Economic development
- Demographic trends
- Government policies on natural resource management
- Market incentives for resource conversion
- Resource accessibility
- Property rights
- Rules and sanctions (enforcement, operational, monitoring)
- Social capital
- Conflicts

The variables were coded as present (1) or absent (0). The data were analyzed using qualitative comparative analysis (QCA) which attempts to bridge the gap between qualitative case-oriented work in social sciences and quantitative analyses (Ragin, 2000, 2008). QCA is especially suited to work with multiple case studies and has been applied to the evaluation of conservation interventions on deforestation using a meta-analysis perspective (Rudel, 2008; Porter-Bolland et al., 2012). QCA uses Boolean algebra to construct an output matrix called the "truth table," which describes several combinations of causal conditions of the explanatory variables that explain the outcome in the dependent variable. The presence or absence of the explanatory variables is related to the outcome through set theory by evaluating the membership of the variables to the set that explains the outcome. We employed the Quine–McCluskey algorithm and the crisp truth table algorithm to perform the Boolean minimization. A frequency cutoff of 1 and a consistency cutoff less than 1 were employed (Ragin, 2008).

Negative biological outcomes in community-managed areas

Negative biological outcomes were significantly associated with rapid economic development, population pressures, market incentives for conversion, conflicts, and accessibility (Figure 10.1). Powerful economic forces and the growing monetization of livelihoods impose considerable pressures on community-managed natural resources (Weinstein & Moegenburg, 2004; Garcia & Pascal, 2005; Armenteras et al., 2009). Within this context, even robust social institutions banking on a traditional conservation ethic may not be a guarantee against ecosystem change. For instance, Peralta & Mather (2000) found major changes in an extractive forest landscape over a 15-year period. Before 1975, the region was mainly exploited by family groups who formed an extractive economy. By 1985, the economy had changed, and people had become dependent on a settled farming economy which resulted in increased deforestation (Peralta & Mather, 2000). In eastern Brazilian Amazonia, demand-led intensification of palm management resulted in the conversion of native floodplain forests into açaí palm-dominated forests (Weinstein & Moegenburg, 2004). Further, the functioning of customary management institutions may not be resilient to population growth and colonization pressures (Bray et al., 2008). In Guatemala and Mexico, poorly governed protected areas and

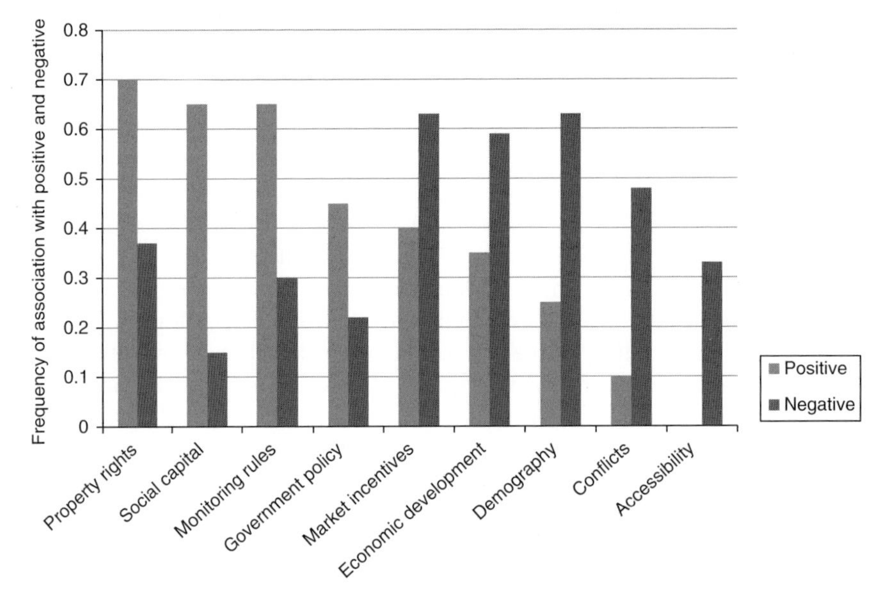

Figure 10.1 **Frequency of association (number of case studies) of selected variables from Ostrom's socioecological systems framework (Ostrom, 2007) with positive and negative biological outcomes. Source: Data from Ostrom (2007)**

community forests were equally ineffective along active colonization fronts. Often, the economic value of the resource influences the rate of natural resource exploitation. The high market value for palm leaves was mostly responsible for unsustainable levels of harvest in Mexico (Endress et al., 2004), and the growing demand for and commercial value of bushmeat was associated with unsustainable levels of hunting in Cameroon (Wilcox & Nambu, 2007). In the Colombian Guyana Shield, the strong influence of illegal coca cultivation on small indigenous territories driven by the lucrative trade in coca resulted in widespread deforestation (Armenteras et al., 2009).

Physical accessibility through roads and the presence of human-constructed facilities such as processing factories play an important role in intensifying extraction levels and ultimately lead to overexploitation. In the Colombian Guyana Shield, roads were a significant factor associated with the expansion of coca cultivation and the degradation of indigenous lands (Armenteras et al., 2009). In northern Mexico, deforestation is driven by resource-specific characteristics, such as location and soil productivity (Perez-Verdin et al., 2009). In another case, palmito factories facilitated the intensification of açaí palm extraction from extractive reserves in eastern Amazonia. In a common property pine forest in Guatemala, Holder (2004) reported deforestation and reduction in forest cover over a 40-year period with some areas protected because of location on steep slopes, but overall, proximity to roads made many areas vulnerable to overexploitation. Increased forest access through a road was associated with hunting-driven depletion of spider monkeys by Huaorani communities in Ecuador (Franzen, 2006). Conflict in the form of social unrest and displacement of human populations (Songer et al., 2009) or between villages over the utilization of resources (Holder, 2004) is also often associated with negative biological outcomes. In other cases, damaging government policies have been largely responsible for negative biological outcomes. Pedlowski et al. (2005) explain that the population rush to Rondonia in Brazil which began in the late 1960s was largely responsible for the massive deforestation caused by the abandonment of colonization projects along the Trans-Amazonian Highway and by changes in land tenure patterns in mid-southern Brazil (Pedlowski et al., 2005). In Myanmar, government policies promoted expansion of sugarcane plantations that required fuelwood for processing and alterations in land registration policies that moved land under the control of large agribusiness leading to deforestation of community-managed areas in a protected area buffer zone (Eberhardt, 2003).

Positive biological outcomes in community-managed areas

A clear trend that emerged in the analyses of positive biological outcomes in community-managed areas was the importance of institutional factors such as strong property rights systems and monitoring and enforcement of harvesting rules, sanctions, and restrictions (Pelkey et al., 2000). Clear property rights were core determinants of the highest proportion of stable forest area being found in community-managed forests in Nepal (Nagendra, 2007) and sustainable Brazil nut extraction in Brazil (Wallgren et al., 2009). In Brazil, indigenous lands that successfully inhibited deforestation within the active agricultural frontier were often inhabited by tribes who actively enforced legal restrictions on natural resource exploitation by outsiders (Nepstad et al., 2006). In the Androy region of Madagascar, social taboos combined with robust property rights and regulation of human behavior through strong sanctions contributed to the continued existence of unique forest habitats (Tengo et al., 2007). In Bhutan, clear operational rules for individual and collective "sokshings"[2] and for equitable appropriation together with traditional institutions that regulated subsistence extraction from sokshing and nonsokshing forests were responsible for successful community management of leaf litter forests in Bhutan (Dorji et al., 2003).

Community governance of mangrove forests in Thailand characterized by well-defined user groups, clear harvesting regulations, strict sanctioning rules, high social capital, strong leadership, and conflict resolution potential has been shown to be more effective for mangrove conservation in comparison with state-controlled forests (Sudtongkong & Webb, 2008). In Brazilian Amazonia, indigenous groups such as the Kayapo have halted the expansion of the agricultural frontier on their lands but allow selective logging and mining. Robust tenurial security and collective choice rules, high social capital, strict surveillance, and ability to negotiate deals with strong economic factors were found to be relevant factors in the expansion of protected lands within a context of absent institutions. Strong self-organization skills enabled the Kayapo to invest economic returns from logging to protect their lands (Schwartzman & Zimmerman, 2005). However, the sustainability of hunting at the landscape scale was largely dependent on source–sink dynamics and the presence of large surrounding tracts of unharvested primary forests. Supportive government resource policies are also strongly associated with positive biological outcomes. For example, the designation of areas for sustainable low-impact farming supports important populations of threatened bird species in the Tonle Sap grasslands of Cambodia (Gray et al., 2007). Strong procommunity forestry policies and laws have enabled positive outcomes for biodiversity in Mexico's community forest enterprises (Bray et al., 2003).

Our analysis highlights the complex, multidimensional aspects of elements underlying biological outcomes in community-managed areas. Taking just one thread in this tangled web of relevant influences, the strong association of economic factors with negative biological outcomes in community-managed areas underscores the need to unravel relationships between economic development, poverty alleviation, and biodiversity outcomes of protected areas. In parallel, the significance of social and institutional influences in positive biological outcomes makes it imperative to simultaneously integrate these considerations into the planning and evaluation of biodiversity outcomes in community-managed areas

(Sunderland et al., 2008). Nonetheless, using diagnostics within a pluralistic framework will be essential to examine processes at multiple levels that ultimately determine the biological effectiveness of CCAs (Berkes, 2007; Ostrom & Cox, 2010).

Biodiversity conservation in community-managed areas: Synergies and trade-offs

To assess the possible contribution of CCAs in achieving relevant Aichi targets, we must analyze their contribution to biodiversity conservation and the nature of their complementarity with SPAs. Specifically, understanding the drivers of synergistic outcomes for biodiversity and economic development is critically important in helping incorporate CCAs into protected area systems. Given that economic considerations have a powerful influence on the effectiveness of CCAs in conserving biodiversity, case studies that examine both biological and economic aspects can provide valuable insights on when there will be synergistic outcomes and when trade-offs will be necessary. Positive livelihood benefits occurred alongside positive biological outcomes in a few examples in our analysis (Bray et al., 2003; Campbell et al., 2007; Leisher et al., 2012). In addition, two site-based examples of synergistic outcomes for biodiversity conservation and poverty alleviation provide useful directions. The Community Markets for Conservation (COMACO) project in the Luangwa Valley, Zambia, represents an example in which biodiversity conservation is associated with sustainable improvements in livelihood and food production (Lewis et al., 2011). With biodiversity conservation as the goal, COMACO's adoption of a local business-based approach allowed the decoupling of economic development from wildlife revenues. This resulted in economic improvements and stabilization of species populations. In Ghana's Wechiau Community Hippo Sanctuary, traditional taboos against the killing of hippopotami were central to an initiative that both protected biodiversity and alleviated severe poverty within a context of a bushmeat crisis and complex ethnic diversity (Sheppard et al., 2010). Key factors associated with the positive outcomes for livelihoods and biological diversity include the existence of strong, traditional taboos, successful economic diversification, improved social capital, true empowerment, equitable distribution of benefits, and strong local governance capacity. Economic forces had been creatively harnessed to uncouple human dependence on natural resources within protected areas. In other cases, economic forces have been successfully coupled with natural areas. Successful ecotourism based on recovering wildlife populations has played an important role in economic empowerment in community conservancies in Namibia (Hoole, 2010). Covering 17% of Namibia's land area, the extensive system of communally managed conservancies complements the ecosystem and biodiversity benefits of the state-managed protected area system. A strong national policy and legislative framework forms the backbone of this system that is widely acknowledged as uniquely successful in integrating conservation benefits with economic empowerment (Bandyopadhyay et al., 2010). In order to make broad and meaningful generalizations, there is a clear need for a rigorous, systematic approach to comparative analyses involving a much larger number of case studies that simultaneously consider the many dimensions that influence biological outcomes (Sunderland et al., 2008).

Nonetheless, while synergistic outcomes for biodiversity conservation and economic development are possible, they are frequently difficult to realize in practice, and the norm is trade-offs and hard choices when it comes to optimizing conservation and development outcomes (McShane et al., 2011). Furthermore, there are suggestions that "win–win" efforts to protect ecosystems and alleviate poverty may be possible only when stakeholders are satisfied with low levels of each outcome, but trade-offs exist when more of either outcome is desired (Ferraro et al., 2011). For example, there is evidence to suggest that protected area systems in Thailand and Costa Rica reduced deforestation and alleviated poverty and socioeconomic gains were likely explained by increased tourism in and around SPAs (Sims, 2010). However, environmental and social impacts were found to be spatially heterogeneous. The spatial characteristics associated with the most avoided deforestation were the characteristics associated with the least poverty alleviation in the Thailand and Costa Rican examples.

Acknowledging the inevitability of trade-offs between biodiversity conservation and human well-being will be fundamental to describing the role of CCAs in biodiversity conservation. Our analyses, while still preliminary, indicate that there are at least two broad generalizations governing biological conservation in CCAs:

1. **Altered community composition:** Many community-managed areas recognized as CCAs are by-products of traditional management of ecosystems for livelihoods, and therefore, rule enforcement in CCAs is likely to be limited to species and communities that are of immediate use to local people. For instance, trees with utilitarian and medicinal values were more abundant in sacred groves than in wildlife reserves (Bhagwat et al., 2005). There is much evidence to suggest altered community composition in CCAs compared to SPAs with observed declines in plant and animal indicator species.
2. **Degradation over time:** Many CCAs especially sacred groves tend to degrade over time either due to their small size or their inability to resist the impacts of intensifying threats (Weinstein & Moegenburg, 2004). Reports of declines in species diversity and forest cover are indicative of the potential long-term limitations of CCAs in conserving biodiversity.

Based on our preliminary analysis and acknowledging the critical need for more systematic comparative analyses including a larger sample size, we identify a set of key guiding principles that could potentially influence the integration of CCAs into national protected area systems (see also McShane et al., 2011):

- **Monitoring and rule enforcement:** The generation of clear rules and restrictions on resource use by strong community institutions, monitoring of adherence to rules, and the enforcement of sanctions and penalties are strongly associated with positive biological outcomes in CCAs (Ostrom & Nagendra, 2006; Chhatre & Agrawal, 2008). The enforcement of laws and regulations concerning encroachment or illegal activities that govern resource exploitation are imperative to prevent degradation and overexploitation of resources in CCAs as in SPAs.
- **Scale of threat:** Robust community institutions can potentially contribute to biological conservation when small-scale threats to biodiversity such as overexploitation can be effectively tackled through strict law enforcement. It is less likely that community-based institutions will continue to be effective in the face of strong economic forces associated with large-scale land conversion. For example, many of the negative biological outcomes in community-managed areas were linked to pressures rising from strong economic forces such as land conversion to lucrative and highly commercialized commodities (e.g., coca, açaí palm, coffee) (Weinstein & Moegenburg, 2004; Garcia & Pascal, 2005).
- **The economic dimension:** Strong economic forces are often directly responsible for the depletion of biodiversity. Yet, on the other hand, poverty alleviation and the pursuit of economic growth frequently have overarching prerogative in CCAs and need to be explicitly considered in conservation planning. Creatively harnessing economic forces to alleviate poverty with tight linkages to biodiversity conservation will contribute to synergistic outcomes for biodiversity and human (Bandyopadhyay et al., 2010; Clements et al., 2010).
- **Policy and legal framework:** Strengthening land and resource tenure through formal recognition is an important incentive to encourage CCAs to join national protected area systems (Berkes, 2007, 2009). Furthermore, government recognition mechanisms need to be backed by transparent legal reforms which clearly indicate the costs and benefits of participation (Robson & Berkes, 2010). The success of communal conservancies in Namibia is heavily influenced by a supportive legislative policy framework (Hoole, 2010).
- **Landscape-scale management:** Protected areas are often too small, and globally, protected area systems are incomplete in coverage to adequately protect biodiversity (Brooks et al., 2009; Butchart et al., 2012). Targeted expansion of protected area systems could include CCAs with high biodiversity values. Overall, landscape-scale conservation strategies that embrace a mix of tenurial regimes (state- and community-managed areas) embedded in large conservation landscapes could significantly help advance biodiversity goals.

Conclusions

The effectiveness of protected area governance in conserving biological diversity is a complex problem, yet one that is highly pertinent to achieving globally agreed-upon conservation goals. Community institutions are an important layer in a multilevel set of processes that underlie variable effectiveness of CCAs in conserving biodiversity targets. Acknowledging that blueprint approaches don't exist and taking a pluralistic approach to defining the framework of conditions that influence the performance of CCAs will be a crucial first step toward integrating them into national protected area systems. Further, considering protected areas as multifaceted SESs with complex vertical and horizontal interplay of institutions will be central to the effectiveness of conservation planning processes that include a diversity of governance types (Berkes, 2010).

Notes

1. The Van Panchayats forests in the midaltitudes of the Indian Himalayas have a long history of autonomous forest management. Villagers have legal tenure over such forests and manage them to fulfill usufruct needs.
2. A sokshing is a small, intensively managed forest plot belonging to an individual, household, or a community for the purpose of producing leaf litter and less frequently fuelwood.

References

Agrawal, A. & Gibson, C. C. (1999). Enchantment and disenchantment: The role of community in natural resource conservation. *World Development*, 27, 629–649.

Armenteras, D., Rodriguez, N., & Retana, J. (2009). Are conservation strategies effective in avoiding the deforestation of the Columbian Guyana Shield. *Biological Conservation*, 142, 1411–1419.

Bajracharya, S. B., Furley, P. A., & Newton, A. C. (2005). Effectiveness of community involvement in delivering conservation benefits to the Annapurna Conservation Area, Nepal. *Environmental Conservation*, 32, 239–247.

Bandyopadhyay, S., Guzman, J. C., & Lendelvo, S. (2010). *Communal conservancies and household welfare in Namibia*. Windhoek: DEA.

Berkes, F. (2007). Community based conservation in a globalized world. *Proceedings of the National Academy of Sciences of the Unites States of America*, 104, 15188–15193.

Berkes, F. (2009). Community conserved areas: Policy issues in historic and contemporary context. *Conservation Letters*, 37, 19–27.

Berkes, F. (2010). Devolution of environment and resource governance: Trends and future. *Environment Conservation*, 37, 489–500.

Bertzky, B., Corrigan, C., Kemsey, J., Kenney, S., Ravilious, C., Besancon, C., et al. (2012). *Protected planet report 2012: Tracking progress towards global targets for protected areas*. Gland & Cambridge: IUCN & UNEP-WCMC.

Bhagwat, S., Kushalappa, C., Williams, P., & Brown, N. (2005). The role of informal protected areas in maintaining biodiversity in the Western Ghats of India. *Ecology and Society*, 10 (1), 8.

Borrini-Feyerabend, G., Kothari, A., & Oviedo, G. (2004). *Indigenous and local communities and protected areas: Towards equity and enhanced conservation.* Gland: IUCN.

Bossart, J. L., Opuni-Frimpong, E., Kuudaar, S., & Nkrumah, E. (2006). Richness, abundance and complementarity of fruit feeding butterfly species in relict sacred forests and forest reserves in Ghana. *Biodiversity and Conservation*, 15, 333–359.

Bray, D. B., Merino-Perez, L., Negreros-Castillo, P., Segura-Warnholtz, G., & Toerres-Rojo Vester, H. F. (2003). Mexico's community managed forests as a global model for sustainable landscapes. *Conservation Biology*, 17, 672–677.

Bray, D. B., Duran, E., Ramos, V. H., Mas, J. F., Velazquez, A., McNab, R. B., et al. (2008). Tropical deforestation, community forests and protected areas in the Maya Forest. *Ecology and Society*, 13, 56.

Brockington, D. (2003). Injustice and conservation: Is 'local support' necessary for sustainable conservation. *Policy Matters*, 12, 22–30.

Brooks, T. M., Wright, S. J., & Sheil, D. (2009). Evaluating the success of conservation actions in safeguarding tropical forest biodiversity. *Conservation Biology*, 23, 1448–1457.

Butchart, S. H., Walpol, M., Collen, B., van Strien, A., Scharlemann, J. P., Almond, R. E., et al. (2010). Global biodiversity: Indicators of recent declines. *Science*, 328 (5982), 1164–1168.

Butchart, S. H., Scharlemann, J. P., Evans, M. I., Quader, S., Arico, S., Arinaitwe, J., et al. (2012). Protecting important sites for biodiversity contributes to meeting global conservation targets. *PLoS ONE*, 7 (3), e32529.

Byers, B. A., Cunliffe, R. N., & Hudak, A. T. (2001). Linking the conservation of culture and nature: A case study of sacred forests in Zimbabwe. *Human Ecology*, 29, 187–218.

Campbell, M. O. (2004). Traditional forest protection and woodlots in the coastal savannah of Ghana. *Environmental Conservation*, 34, 225–232.

Campbell, L. M., Haalboom, B. J., & Trow, J. (2007). Sustainability of community based conservation: Sea turtle egg harvesting in Ostional (Costa Rica) ten years later. *Environmental Conservation*, 34, 122–131.

Carillo, E., Wong, G., & Cuaron, A. D. (2000). Monitoring mammal populations in Costa Rican protected areas under different hunting restrictions. *Conservation Biology*, 14, 1580–1591.

CBD. (2010). *Decisions adopted by the CoP to the CBD at its 10th meeting (UNEP/CBD/COP/DEC/X/2).* Montreal: Secretariat of the CBD.

Chhatre, A. & Agrawal, A. (2008). Forest commons and local enforcement. *Proceedings of the National Academy of Sciences of the United States of America*, 105, 13286–13291.

Clements, T., John, A., Nielsen, K., An, D., Tan, S., & Milner-Gulland, E. J. (2010). Payments for biodiversity conservation in the context of weak institutions: Comparison of three programs from Cambodia. *Ecological Economics*, 69, 1283–1291.

Dalle, S. P., de Blois, S., Caballero, J., & Johns, T. (2006). Integrating analysis of local land use regulations, cultural perceptions and land use/land cover data for assessing the success of community based conservation. *Forest Ecology and Management*, 222, 370–383.

Dorji, L., Webb, E. L., & Shivakoti, G. P. (2003). Can a nationalized forest management system uphold local institutions? The case of leaf little forest (Sokshing) management in Bhutan. *Asian Studies Review*, 27, 341–359.

Eberhardt, K. (2003). A review of challenges to sustainable development in the uplands of Myanmar. In X. J. Mikesell (Ed.), *Landscapes of diversity: Indigenous knowledge, sustainable livelihoods and resource governance in montane mainland Southeast Asia* (pp. 101–111). Kunming: Yunnan Science & Technology.

Endress, B. A., Gorchov, D. L., Peterson, M. B., & Serrano, E. P. (2004). Harvest of palm (Chamaedorea radicalis), its effects on leaf production and implications for sustainable management. *Conservation Biology*, 18, 822–831.

Fabricius, C., Burger, M., & Hockey, P. A. (2003). Comparing biodiversity between protected areas and adjacent rangeland in Xeric succulent thicket, South Africa: Arthropods and reptiles. *Journal of Applied Ecology*, 40, 392–403.

Ferraro, P. J., Hanauer, M. M., & Sims, K. R. (2011). Conditions associated with protected area success in conservation and poverty reduction. *Proceedings of the National Academy of Sciences of the United States of America*, 108, 13913–13918.

Franzen, M. (2006). Evaluating the sustainability of hunting: A comparison of harvest profiles across the three Huaorani communities. *Environmental Conservation*, 33, 36–45.

Garcia, C. A. & Pascal, J. P. (2005). Sacred forests of Kodagu: Ecological value and social role. In G. Cederlof & K. Sivaramakrishnan (Eds.), *Ecological nationalisms: Nature, livelihoods and identities in South Africa* (pp. 199–232). Seattle: University of Washington Press.

Gardner, T. A., Caro, T., Fitzherbert, E. B., Banda, T., & Lalbhai, P. (2007). Conservation value of multiple use areas in East Africa. *Conservation Biology*, 21, 1516–1525.

Gray, T. N., Chamnan, H., Borey, R., Collar, N. J., & Dolman, P. M. (2007). Habitat preferences of a globally threatened bustard provides support for community based conservation in Cambodia. *Biological Conservation*, 138 (3), 341–350.

Holder, C. D. (2004). Changes in structure and cover of a common property pine forest in Guatemala. *Environmental Conservation*, 31, 22–29.

Hoole, A. F. (2010). Place power prognosis: Community based conservation, partnerships and ecotourism enterprise in Namibia. *International Journal of the Commons*, 4, 78–99.

Johnson, K. A. & Nelson, K. C. (2004). Common property and conservation: The potential for effective communal forest management within a national park in Mexico. *Human Ecology*, 32, 703–733.

Leisher, C., Hess, S., Boucher, T. M., van Beukering, P., & Sanjayan, M. (2012). Measuring the impacts of community based grasslands management in Mongolia's Gobi. *PLoS ONE*, 7 (2), e30991.

Lewis, M. (2003). Cattle and conservation at Bharatpur: A case study in science and advocacy. *Conservation and Society*, 1, 1–22.

Lewis, D., Bell, S. D., Fay, J., Bothid, K. L., Gatere, L., Kabilaa, M., et al. (2011). Community markets for conservation (COMACO) link biodiversity conservation with sustainable improvements in livelihoods and food production. *Proceedings of the National Academy of Sciences of the United States of America*, 108, 13957–13962.

McShane, T. O., Hirsch, P. D., Trung, T. C., Songorwa, A. N., Kinzig, A., Monteferri, B., et al. (2011). Conservation and development in tropical forest landscapes: A time to face the tradeoffs. *Biological Conservation*, 144, 966–972.

Nagendra, H. (2007). Drivers of reforestation in human dominated forests. *Proceedings of the National Academy of Sciences of the United States of America*, 104, 15218–15223.

Nawaz, M. (2008). Assessment of avifauna composition at Paya Indah wetland peat swamp forest, Selangor Peninsular Malaysia: International symposium and workshop on tropical peatland. Kuching, Sarawak, Malaysia 19–22 August 2008, Ecology and Biodiversity.

Nepstad, D., Schwatzman, S., Bamberger, B., Santilli, M., Ray, D., Schlesigner, P., et al. (2006). Inhibition of Amazon deforestation and fire by parks and indigenous lands. *Conservation Biology*, 20, 65–73.

Ostrom, E. (1990). *Governing the commons: The evolution of institutions for collective action.* Cambridge: Cambridge University Press.

Ostrom, E. (2007). A diagnostic approach for going beyond panaceas. *Proceedings of the National Academy of Sciences of the Unites States of America*, 104, 15181–15178.

Ostrom, E. & Cox, M. (2010). Moving beyond panaceas: A multi tiered diagnostic approach for social-ecological analysis. *Environmental Conservation*, 37, 451–463.

Ostrom, E. & Nagendra, H. (2006). Insights on linking forests, trees and people from the air, on the ground and in the lab. *Proceedings of the National Academy of Sciences of the Unites States of America*, 103, 19224–19331.

Oviedo, G. (2006). Community conserved areas in South America. *PARKS*, 16, 49–55.

Pathak, N., Bhatt, S., & Balainorwala, T. (2004). *Community conserved areas: A bold frontier for conservation—Briefing Note 5.* Iran: IUCN WCPA-CEESP TILCEPA.

Pedlowski, M. A., Matricardi, E. A., Skole, D., Cameron, S. R., Chomentowski, W., Fernandes, C., et al. (2005). Conservation units: A new deforestation frontier in the Amazonian state of Rondonia, Brazil. *Environmental Conservation*, 32, 149–155.

Pelkey, N. W., Stoner, C. J., & Caro, T. M. (2000). Vegetation in Tanzania: Assessing long term trends and effects of protection using satellite imagery. *Biological Conservation*, 94, 297–309.

Peralta, P. & Mather, P. (2000). An analysis of deforestation patterns in the extractive reserves of Acre, Amazonia from satellite imagery: A landscape ecological approach. *International Journal of Remote Sensing*, 21, 2555–2570.

Peres, C. A. & Nascimento, H. S. (2006). Impact of game hunting by the Kayapo of south eastern Amazonia: Implications for wildlife conservation in tropical forest indigenous reserves. *Biodiversity and Conservation*, 15, 2627–2653.

Perez-Verdin, G., Kim, Y., Hospodarsky, D., & Tecle, A. (2009). Factors driving deforestation in common pool resources in northern Mexico. *Journal of Environmental Management*, 90, 331–340.

Porter-Bolland, L., Ellis, E. A., Guariguata, M. R., Ruiz-Mallen, I., Negrete-Yankelevich, S., & Reyes-Garcia, V. (2012). Community managed forests and forest protected areas: An assessment of their conservation effectiveness across the tropics. *Forest Ecology and Management*, 268, 6–17.

Ragin, C. C. (2000). *Fuzzy set social science.* Chicago: University of Chicago Press.

Ragin, C. C. (2008). *Redesigning social inquiry: Fuzzy sets and beyond.* Chicago: University of Chicago Press.

Rai, N. & Uhl, C. F. (2004). Forest product use, conservation and livelihoods: The case of Uppage fruit harvest in the Western Ghats, India. *Conservation and Society*, 2, 289–313.

Ranganathan, J., Daniels, R. J., Subash Chandran, M. D., Ehrlich, P. R., & Daily, G. C. (2008). Sustaining biodiversity in ancient tropical countryside. *Proceedings of the National Academy of Sciences of the United States of America*, 105, 17852–17854.

Robson, L. W. & Berkes, F. (2010). Applying resilience thinking to questions of policy for pastoralist systems: Lessons from the Gabra of northern Kenya. *Human Ecology*, 38, 335–350.

Rudel, T. K. (2008). Meta analyses of case studies: A method for studying regional and global environmental change. *Global Environmental Change*, 18, 18–25.

Ruiz-Perez, M., Almeida, M., Dewi, S., Costa, E. M., Pantoja, M. C., Puntodewo, A., et al. (2005). Conservation and development in Amazonian extractive reserves: The case of Alto Jurua. *Ambio*, 34, 218–223.

Saha, S. & Howe, H. F. (2003). Species composition and fire in a dry deciduous forest. *Ecology*, 84, 3118–3123.

Saha, S. & Howe, H. F. (2006). Stature of juvenile trees in response to anthropogenic fires in a tropical deciduous forest of central India. *Conservation and Society*, 4, 619–627.

Sarkar, R. (2008). Decentralized forest governance in central Himalayas: A re-evaluation of outcomes. *Economic and Political Weekly*, 43, 54–63.

Schwartzman, S. & Zimmerman, B. (2005). Conservation alliances with indigenous peoples of the Amazon. *Conservation Biology*, 19, 721–723.

Setsaas, T. H., Holmern, T., Mwakalebe, G., Stokkec, S., & Roskraft, E. (2007). How does human exploitation affect impala populations in protected and partially protected areas? A case study from the Serengeti Ecosystem. *Biological Conservation*, 136, 563–570.

Shackleton, C. M. (2000). Comparison of plant diversity in protected and communal lands in the Bushbuckridge Lowveld savannah, South Africa. *Biological Conservation*, 94, 273–285.

Shahabuddin, G. & Rao, M. (2010). Do community conserved areas effectively conserve biological diversity? Global insights and the Indian context. *Biological Conservation*, 143, 2926–2936.

Sheppard, D. J., Moehrenschlager, A., McPherson, J. M., & Mason, J. J. (2010). Ten years of adaptive community governed conservation: Evaluating biodiversity protection and poverty alleviation in a West African hippopotamus reserve. *Environmental Conservation*, 37, 270–282.

Sims, K. R. (2010). Conservation and development: Evidence from Thai protected areas. *Journal of Environmental Economics and Management*, 60, 94–114.

Somanathan, E., Prabhakar, R., & Mehta, B. S. (2009). Decentralization for cost effective conservation. *Proceedings of the National Academy of Sciences of the United States of America*, 106, 4143–4147.

Songer, M. E., Aung, M., Senior, D., DeFries, R., & Leimgruber, P. (2009). Spatial and temporal deforestation dynamics in protected and unprotected dry forests: A case study from Myanmar (Burma). *Biodiversity Conservation*, 18, 1001–1018.

Sudtongkong, C. & Webb, E. J. (2008). Outcomes of state versus community based mangrove management in southern Thailand. *Ecology and Society*, 13, 27.

Sunderland, T. C., Ehringhaus, C., & Campbell, B. M. (2008). Conservation and development in tropical forest landscapes: A time to face the tradeoffs. *Environmental Conservation*, 34, 276–279.

Tengo, M., Johansson, K., Rakotondrasoa, F., Lundberg, J., Andriamaherilala, J., Rakotoarisoa, J., et al. (2007). Taboos and forest governance: Informal protection of Hot Spot dry forest in southern Madagascar. *Ambio*, 36, 683–691.

Tucker, C. M. (2004). Community institutions and forest management in Mexico's butterfly reserve. *Society and Natural Resources*, 17, 569–587.

Wallgren, M., Skapre, C., Bergstrom, R., Danell, K., Bergstrom, A., Jakobsson, T., et al. (2009). Influence of land use on the abundance of wildlife and livestock in the Kalahari, Botswana. *Journal of Arid Environments*, 73, 314–321.

Webb, E. L. & Gautam, A. P. (2001). Effects of community forest management on the structure of diversity of a successional broad leaf forest in Nepal. *International Forestry Review*, 3, 146–157.
Weinstein, S. & Moegenburg, S. (2004). Acai palm management in the Amazon Estuary: Course for conservation or passage to plantation. *Conservation and Society*, 2, 315–346.
Wilcox, A. S. & Nambu, D. M. (2007). Wildlife hunting practices and bushmeat dynamics of the Banyangi and Mbo people of south western Cameroon. *Biological Conservation*, 134, 251–261.

11

The Importance of Asia's Protected Areas for Safeguarding Commercially High Value Species

J. Walston, E. J. Stokes and S. Hedges

Wildlife Conservation Society, Bronx, NY, USA

Introduction

The internationally illegal wildlife trade directly threatens the survival of commercially high value (CHV) species. CHV species are those for which there is a significant trade in body parts or whole animals (dead or alive) and the driver of the trade is consumer demand largely in international markets. Occasionally, national markets contribute to this demand but rarely, if ever, are local markets significant. To meet this demand, many CHV species are illegally hunted or harvested specifically so that they or their parts can be traded. Examples of these CHV species include some timber species and orchids, as well as vertebrate species such as turtles, pangolins, wild cattle, tigers, elephants and rhinoceroses.

Consumer demand for wildlife products is extensive and deeply rooted in East Asian culture. This demand is growing as a result of: the rapidly rising disposable incomes of middle classes in Asia, especially in China (Van Song, 2008; Martin & Vigne, 2011; CITES, 2013; Vandegrift, 2013); the changing patterns of wildlife product use such as the increased use of rhinoceros horn in nontraditional preparations such as hangover cures and even aphrodisiacs (Milliken & Shaw, 2012); and the increasing influence of speculators (Mason et al., 2012; 't Sas-Rolfes & Fitzgerald, 2013). The internationally illegal trade in wildlife is now worth between US$5- and US$20- billion per year (Rosen & Smith, 2010), and it increasingly involves highly organized and sophisticated criminal networks (Rademeyer, 2012; CITES, 2013; Hubschle, 2014). According to some estimates,

Protected Areas: Are They Safeguarding Biodiversity?, First Edition.
Edited by Lucas N. Joppa, Jonathan E. M. Baillie and John G. Robinson.

wildlife trade is now the fifth largest international criminal activity after narcotics, counterfeiting, and the illicit trafficking of people and of oil (Haken, 2011; Dudley et al., 2013).

Given the scale and nature of the threat posed by the international illegal wildlife trade, CHV species require radically enhanced protection from illegal hunting as well as protection of their habitat in order for functioning populations to survive. Walston et al. (2010) argued that for "financially valuable species like the tiger, intensive protection is paramount" and there is a need to focus efforts on those places with the best chances of success, mostly in already protected areas. In this chapter, we develop and extend the analysis of Walston et al. (2010) by considering its relevance to other CHV species in Asia. We present evidence to suggest that protected areas have been disproportionately more successful in stemming declines in CHV species compared to other land uses in Asia but that the protected area system is still limited and too frequently its management is not effective. Finally, we discuss the need both invest more heavily in protecting existing protected areas and extend and connect them, and to establish new protected areas where sufficient protection measures can be put in place.

In this chapter, we focus in this chapter on tiger (*Panthera tigris*), Asian elephant (*Elephas maximus*), Asian rhinoceros species (greater one-horned rhinoceros, *Rhinoceros unicornis*; Javan rhinoceros, *Rhinoceros sondaicus*; Sumatran rhinoceros, *Dicerorhinus sumatrensis*), and two Asian wild cattle species (banteng, *Bos javanicus*; gaur, *Bos gaurus*). We chose these species because a relatively large amount of data has been amassed on them, allowing us to draw more robust conclusions than those for species such as pangolin. We conclude that these species provide illustrative and valuable examples of the importance of protected areas in the safeguarding of CHV species in Asia and even beyond. For example, much of what we say applies to African rhinoceros species, *Ceratotherium simum* and *Diceros bicornis*, and increasingly to African elephants, *Loxodonta africana*. While all these species are large-bodied mammals and are not taxonomically representative of all CHV species, their commercial value is the most obvious common feature. We therefore suggest that the arguments presented hold for a wide variety of species, regardless of their body size or taxonomic position.

For tigers, the main trade is in body parts (bones, skin, and other parts) and is international (Mills & Jackson, 1994; Moyle, 2009; Tilson & Nyhus, 2010). For Asian rhinoceroses, the main trade is in horns and again is primarily international (Milliken et al., 2009; Martin, 2010). The main commercial trade-driven threat to wild cattle in Asia is the international market for trophies (horns usually attached to partial skulls); there is also commercial trade in meat, but that is mainly national (Srikosamatara et al., 1992; Srikosamatara & Suteethorn, 1994; Duckworth & Hedges, 1998; Pedrono et al., 2009). For Asian elephants, the trade is in ivory and other body parts but also in live elephants and is also primarily international in nature (Sukumar, 2003; Hedges et al., 2006; Shepherd & Nijman, 2008; Stiles, 2009).

Methods

Range and population numbers

We assembled the most recent and authoritative range maps of the seven target species both of contemporary and, where available, historic ranges. We also compiled all available data on population numbers (see data sources in the following). Range maps and population estimates were used to assess what proportion of the species' ranges and populations are now within protected areas and thus to what extent they are dependent on protected areas. We then looked at overlap in the species' ranges to better assess the costs of conserving CHV species given that many actions aimed at conserving one CHV species will benefit other such species.

Data sources

We evaluated data sources for each of the seven species to determine the most current and accurate data available. For tigers, we used data compiled by Sanderson et al. (2006) for their Tiger Conservation Landscape (TCL) analysis and Walston et al. (2010) for their source site[1] analysis. Specifically, we defined tiger range as all TCLs that contain a source site or potential source site, except where our knowledge of tiger distribution suggested the need for changes based on, for example, recent surveys (Sanderson et al., 2006; Walston et al., 2010).

For Asian elephants, we used data compiled at a Wildlife Conservation Society-led Asian elephant mapping workshop organized jointly with the IUCN SSC Asian Elephant Specialist Group (AsESG) and WWF (Hedges et al., 2009).

For gaur and banteng in Southeast Asia, we used data on the species' range and population sizes compiled and published by the IUCN SSC Asian Wild Cattle Specialist Group (IUCN, 2010); for gaur in South Asia, we used the data compiled for the Global Mammal Assessment/IUCN Red List (Duckworth et al., 2008). For gaur, banteng, and Asian elephants, we used only the "confirmed" and "possible" range categories; in essence, these categories include only those areas known to contain the species or for which there is relatively little doubt that they still occur, respectively.

For Asian rhinoceroses, we used data on range and population numbers included in recent reports from the IUCN SSC Asian Rhino Specialist Group (AsRSG) and in particular Talukdar et al. (2010) and Emslie et al. (2013), as well as Goossens et al. (2013), Subedi et al. (2013), and Thapa et al. (2013).

For protected areas, we used the World Database on Protected Areas (WDPA, 2010) making modifications where we knew protected area boundaries or designated categories to be incorrect. For the purposes of this analysis, we considered only protected areas in IUCN categories I–IV (Dudley, 2008). All spatial analyses were conducted using standard GIS software (ESRI, 2011).

Costs of conserving CHV species

To assess the costs of conserving CHV species, we used the annual costs of protecting and monitoring tigers and their prey as presented by Walston et al. (2010). These estimated costs were obtained, where possible, from those responsible for managing sites such as protected area authorities and agencies supporting that management, supplemented by published national government figures. In addition to protection, management, and monitoring activities, site personnel were also encouraged to include additional activities which they considered essential. A number of sites, therefore, included: community engagement; mitigation of human wildlife conflict, intelligence gathering informant networks, and local monitoring of trade within their estimates of financial needs. Historical conservation infrastructure development and costs related to the relocation of communities within sites were not included in the analysis. Costs were weighted with cost of living (salaries) and logistical difficulty for sites. Results were compared to formal data sets (both governmental and global databases) to highlight inconsistencies.

To assess the incremental costs of conserving the six species, in addition to the tiger, we used two approaches:

Firstly, for sites which contained tigers in addition to one or more of the other CHV species, the additional cost of protecting and monitoring those species was calculated as the incremental cost of necessary management beyond that needed to protect and monitor tigers and their prey. For gaur and banteng, the additional costs were assumed to be zero since these are tiger prey species and the costs of protecting and monitoring for tiger prey species were already included in the tiger cost estimates. For Asian elephants and rhinoceroses, we considered the costs of law enforcement to be already included by the law enforcement component for tigers but added the additional costs of species-specific monitoring methods (primarily fecal DNA-based capture–recapture methods for the pachyderm species rather than the primarily camera trap-based methods for tigers) to the baseline monitoring costs estimated by Walston et al. (2010). We based these monitoring costs on our experience of running monitoring programs for these species across Asia and took into account the appropriate cost of manpower in each of the countries (Hedges, 2012; Hedges et al., 2013). For elephants, an additional cost was the site-specific cost of human–elephant conflict (HEC) mitigation. HEC directly affects elephant numbers through retaliatory killing and indirectly affects them through the setting of fires, destruction of protected area infrastructure, and the sabotaging of conservation projects (Hedges, 2006). Again, costs were based on our experience of running HEC mitigation projects across Asia (Hedges & Gunaryadi, 2010; McWilliam et al., 2010; Gunaryadi et al., in review).

Second, for those sites without tigers but with one or more of the other six CHV species, the base costs for protection were calculated using the country-specific average cost per square kilometer as calculated for that country in the Walston et al. (2010) analysis for tigers (given that in most cases, tiger source sites fall within protected areas). Additional cost for species-specific monitoring costs and HEC mitigation were then added as before.

Results

CHV species in Asia are now largely restricted to protected areas

The ranges of all seven CHV species are now characterized by a pattern of fragmentation and isolation (Figure 11.1). All these species now exist in only a small fraction of their historical range. For tigers and Asian elephants, for which the historical range data are most well known, the proportion of their historical ranges that still contain these species is just 7% and less than 10%, respectively. For the Asian rhinoceroses, almost all of their current range and virtually all individuals in all populations are now within protected areas. Moreover, about 87% of all greater one-horned rhinoceroses now occur in just two protected areas (Kaziranga and Chitwan national parks). The restriction to protected areas is not as dramatic for the other four species, but the overall pattern is similar: at least 65% of the total population of tigers, Asian elephants, banteng, and gaur are now within protected areas (Table 11.1).

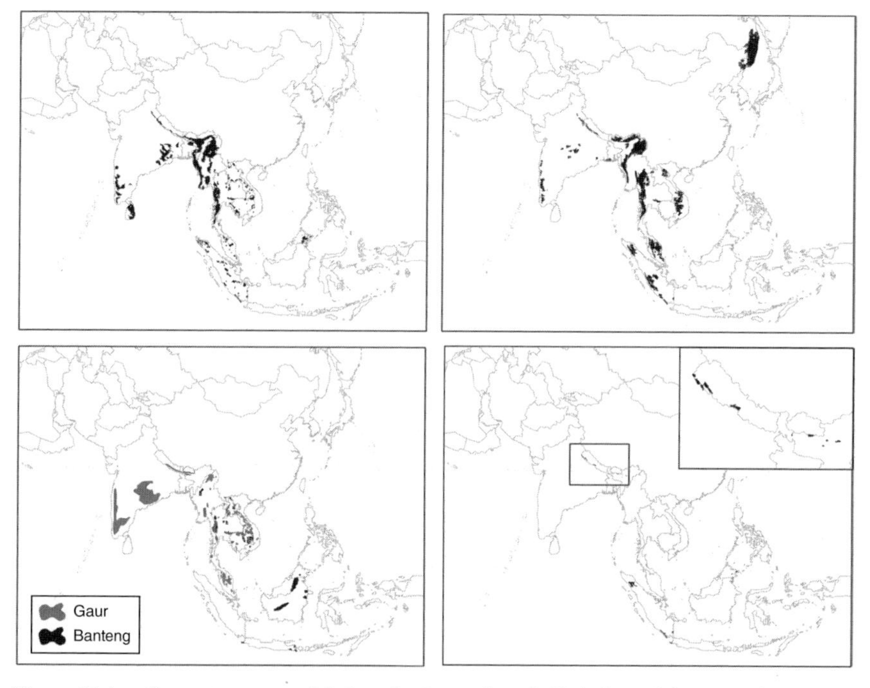

Figure 11.1 **Current ranges of Asian elephants (top left) (adapted from Hedges et al., 2009), tigers (top right) (adapted from Walston et al., 2010), gaur and banteng (bottom left) (adapted from IUCN, 2010), and Asian rhinoceroses (bottom right) (adapted from Talukdar et al., 2010), with Javan rhinoceros solely on the island of Java, Sumatran rhinoceros only the islands of Borneo and Sumatra, and greater one-horned rhinoceros only in India and Nepal (inset)**

Table 11.1 **Proportion of the seven CHV species' range and total population within protected areas**

Species	Range area (km²)	Range in protected areas (km²)	Range in protected areas (%)	Total population in protected areas (%)
Tiger	1,111,101	219,649	20	73–83
Asian elephant	926,776	186,339	20	65–85
Gaur	778,526	147,250	19	65–85
Banteng	190,933	106,180	56	85–95
Greater one-horned rhinoceros	5,384	5,200	97	>98
Sumatran rhinoceros	25,205	24,650	98	>98
Javan rhinoceros	996	996	100	100

There is significant overlap in the remaining range of CHV species in Asia

A large proportion (45.3%; $n = 211$) of the protected areas in the range of the seven CHV species contain more than one species, with many protected areas containing two (21.3%; $n = 99$), three (16.3%; $n = 76$), or four (7.7%; $n = 36$) CHV species (Figure 11.2).

Costs of conserving CHV species

The cost estimates used in the Walston et al. (2010) analysis came from 126 individuals working at 68 sites across Asia. That analysis gave an estimated average cost of protecting and monitoring tigers effectively at all 42 tiger source sites of US$82 million per year or US$930 per km² per year. Of the 42 tiger source sites, 31 contain at least 1 other CHV species. Our extension of that expert opinion-based analysis of the additional costs of conserving these species in source sites indicates that it is approximately 1.0–1.15 times the cost for tigers alone (it is 1.0 if only gaur or banteng are present in addition to tigers). Thus, for the 31 source sites with tigers and at least 1 other CHV species, the cost of protecting and monitoring all CHV species (including tigers) effectively is approximately US$65 million per year compared to US$59 million per year to protect just tigers and their prey in those source sites. The total cost of protecting tigers and the other CHV species in the 42 source sites (i.e., including the 11 source sites that contain only tigers) is US$89 million per year.

For all protected areas with CHV species (i.e., the 42 source sites plus all other protected areas with at least 1 of the 7 CHV species including tigers), the cost of protecting and monitoring all CHV species effectively is approximately US$387 million per year

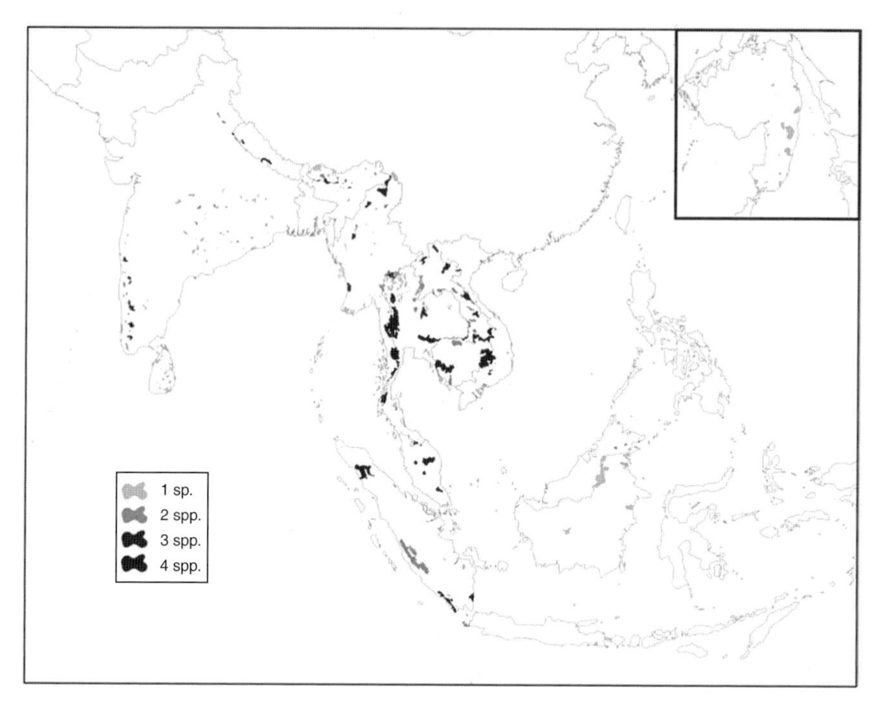

Figure 11.2 **Overlap in the range of CHV species in protected areas showing how many of the seven species occur in each protected area**

(equivalent to US$954 per km^2 per year) or approximately 4.7 times the annual cost of protecting and monitoring tigers (and prey) effectively at just the 42 tiger source sites.

Discussion

Limitations of data quality

This study selected species for which we have substantial data sets on distribution and abundance, as well as knowledge on their international trade. However, even with these relatively well-studied species such as elephants, tigers, and rhinoceroses, their abundance and distribution are still rather poorly known (Duckworth & Hedges, 1998; Blake & Hedges, 2004; Ahmad Zafir et al., 2011). Indeed, only a small fraction of the world's protected areas have wildlife monitoring systems in place (Coad et al., 2013). Even when monitoring work is undertaken, appropriate methods are not always used. Species' range maps are typically simple presence/probable absence maps, based on expert opinion rather than derived using formal methods such as occupancy modeling. Abundance estimates, where they exist, are too often based on untested assumptions or use unreliable methods (Karanth et al., 2003; Blake & Hedges, 2004; Karanth & Nichols, 2010; Hedges, 2012).

That being said, the absence of high-quality data is not overly problematic for this analysis. With species' distributions, the uncertainty is primarily about whether the species occur outside protected areas: if we had restricted ourselves to just "confirmed" range categories rather than also including, as we do, "possible" categories, the proportion of the species range within protected areas would be significantly higher than indicated in our results (Table 11.1), which would only strengthen the arguments we make in this chapter. Secondly, given the lack of, or uncertainty about, abundance estimates for the seven CHV species in many places, we used the best available compilations of expert opinion on the species' abundance and distribution to estimate the likely proportion of the species' total population in protected areas, giving an appropriate range to reflect the uncertainty in the estimates. Thus, our estimates of the total proportions of the seven species' total populations that occur inside protected areas should be relatively robust. Moreover, for the purposes of the argument we present here (i.e., that protected areas in Asia are disproportionately important for the conservation of CHV species), it does not matter whether, for example, 71 or 83% of a given species' total population is in protected areas: as long as the great majority of the species' populations are inside protected areas, which is clearly the case, the argument holds.

The disproportionate importance of the protected area network in Asia is revealed by the distribution of CHV species

The great majority of the remaining global populations of the seven Asian CHV species considered here are now inside protected areas. This disproportionate importance of the protected area network in Asia is not because that network is extensive and well placed. In fact, the Asian protected area network is dominated by relatively small sites (mean size = 916 km^2), and several major vegetation types (e.g., lowland forests and alluvial grasslands) preferred by the CHV species considered here are underrepresented in the network. The reason for this over-representation of CHV species in protected areas is that the protected areas or at least subsets of them (typically the better secured protected areas) have provided a measure of protection from the catastrophic range and population collapses these species have suffered.

That collapse was driven primarily by a combination of illegal hunting and habitat loss, with the relative significance of these two drivers varying among the species and among countries (as well as over time): for Asian rhinoceroses (see Hoogerwerf, 1979; Dinerstein, 2003; Ahmad Zafir et al., 2011; Brook et al., 2012; Thapa et al., 2013); for tigers (see Seidensticker et al., 1999; Tilson & Nyhus, 2010; Walston et al., 2010); for Asian elephants (see Leimgruber et al., 2003; Sukumar, 2003; Hedges, 2006; Gopala et al., 2011); for wild cattle (see Hedges, 1995; Duckworth & Hedges, 1998; Duckworth et al., 2008). Thus, it is critical that the remaining habitat of the CHV species is protected effectively but, as we argue here, habitat protection is not sufficient for CHV species especially given the high rates of illegal hunting driven by the demand for their horns, bones, skin, ivory, and other body parts and the trend in that demand, which shows no sign of abating (Martin & Vigne, 2011; Dudley et al., 2013; Thapa et al., 2013).

Protected areas provide the best hope for the dedicated antipoaching efforts needed to conserve CHV species in Asia

Criteria for effective conservation of CHV species in Asia

The continually escalating rate of illegal hunting of CHV species, which is driven by the international trade in their body parts, is not abated merely by the presence of protected areas. Instead, dedicated law enforcement is required. The paramount importance of such dedicated antipoaching efforts for CHV species is shown clearly by the large areas of suitable habitat across the species' recent ranges that are now largely empty of CHV species because of illegal hunting (and for elephants at least, capture for illegal international trade in live animals). Particularly illustrative cases include rhinoceroses across the Southeast Asian mainland and on Borneo (Ahmad Zafir et al., 2011; Brook et al., 2012), tigers in Indochina (Duckworth & Hedges, 1998; O'Kelly et al., 2012), banteng outside of Java (Hedges, 1995; Pedrono et al., 2009), gaur in Peninsular Malaysia (Duckworth et al., 2008), and elephants in Myanmar (Leimgruber et al., 2008).

It is therefore important to identify those aspects and features of protected areas that are consequential in effectively conserving CHV species. If protected areas are to become more effective and CHV species are to be conserved across broader landscapes including outside of protected areas, then there is obvious merit in defining those relevant features.

Reviewing the nature of the threats to CHV species, the tiger assessment of Walston et al. (2010), and the results of the present analysis for the six other CHV species in Asia as well as our many years of experience working to conserve these species leads us to suggest that effective conservation of CHV species requires the following criteria to be met:

1. A legal framework is in place for the prevention of illegal killing and harvesting of CHV species.
2. The land use type provides a formal and explicit mandate for protecting species of high conservation value.
3. The boundaries of the land use type need to be clear, well marked, and widely recognized.
4. There is a dedicated protection force with the authority for interdiction (prevention, arrest, and general law enforcement).

5. There is sufficient capacity for law enforcement activities (e.g., ranger numbers, training and equipment levels, and ranger effort per month per square kilometer).
6. There is an operational budget for regular protection and monitoring activities (and other species-specific activities as required).

In Asia, typically only protected areas meet or even begin to meet these criteria. While far from perfect, many protected areas in Asia provide a legal mandate and evidence of the political will for protection and other conservation interventions. Protected areas usually have government budgets far in excessive of typical external conservation funding, and they typically have the staff and at least some of the infrastructure necessary for protection and other conservation interventions (Walston et al., 2010). Unlike the situation in many parts of Africa, in Asia, there are not extensive private concessions, community lands, or other land use types that can provide effective refuges for CHV species. Some land use types do have the potential to provide additional areas in which CHV species can be protected (e.g., through legally mandated antipoaching activities); these include well-managed buffer zones of the type seen around some protected areas in India (Bindra & Karanth, 2013), wildlife corridors with community-led antipoaching units as seen in Nepal (Thapa et al., 2013), and hunting leases in Russia (Miquelle et al., 2005). Nevertheless, such land use types are rare in most of Asia. The importance of these other land use types is also likely to be higher if they are adjacent to protected areas as seen clearly, for example, in the Nepalese case referred to earlier.

Intensive protection efforts have been successful

The number of places where protection has been demonstrably successful is small, particularly in terms of population recovery. For tigers, there are some notable successes in Indian protected areas, with the best example being the Nagarahole National Park, which has seen a 400% increase in tiger numbers as a result of well-managed enforcement efforts beginning in the 1970s and has now maintained a high density for 30 years (Karanth, 2002; Karanth et al., 2006). In Southeast Asia, probably the only example of an increase in tiger numbers as a result of well-managed protection efforts in a protected area comes from Huai Kha Khaeng Wildlife Sanctuary in Thailand (unpublished WCS data). For the greater one-horned rhinoceros, there are several well-documented examples of how intensified protection systems and institutional strengthening reversed previous poaching-driven population collapses to achieve very significant growth in numbers. For example, in Chitwan National Park in Nepal, rhinoceros numbers increased from 100 in 1966 to 544 in 2000 (DNPWC, 2000), and there has been a 7.7% annual growth rate in the rhinoceros population in Nepal's protected areas (Chitwan National Park, Bardia National Park, and Suklaphanta Wildlife Reserve) from 435 animals in 2008 to 534 animals in 2011 (Thapa et al., 2013). As Thapa et al. (2013)

note, "Protection is the key to the successful recovery of the greater one-horned rhinoceros in Nepal."

Protection is essential but is not being done well enough in many protected areas in Asia

In many parts of Asia, there are significant areas of CHV species habitat, including inside protected areas that currently do not contain CHV species. This is primarily because of illegal hunting. For example, in some parts of Cambodia, tiger habitat and prey still exist in some protected areas, but these contain no tigers and there is no opportunity for immigration. Reestablishing tigers in most areas in Cambodia would require translocation (O'Kelly et al., 2012). As another example, in the 1990s, the Indonesian island of Java was one of the remaining strongholds for banteng with 6 of the 8 populations known to contain more than 50 individuals and all entirely in protected areas (Hedges, 1995). By the mid-2000s, two of the six populations had been lost, and another was teetering on the brink of local extirpation. This collapse was due primarily to illegal hunting for meat (for national trade and local consumption) in the absence of effective law enforcement in the protected areas (IUCN, 2010). Vietnam recently lost its last rhinoceros to poachers (rendering the subspecies *Rhinoceros sondaicus annamiticus* extinct), and there are fears that the country's banteng may soon be lost to poachers too (Pedrono et al., 2009; Brook et al., 2012). Unfortunately, such examples are far from unusual across much of Asia.

Given (i) the disproportionate importance of Asian protected areas for the seven CHV species discussed in this chapter (and doubtless for many other such species too), (ii) the examples of successful CHV species protection in protected areas discussed earlier, and (iii) the lack of alternative land use types that lend themselves to CHV species conservation in Asia, it is clear that there need to be significant improvements to species protection practice within protected areas. This is further highlighted by the fact that protected areas are becoming ever more important to protect many wildlife species as human populations grow and further wildlife habitat is lost outside protected areas. Furthermore, those threats specific to CHV species, particularly illegal killing driven by the rising demand for their body parts, show no signs of abating.

Conservation interventions are more expensive than is often realized, but the sums required are attainable

The true cost of CHV species conservation cannot be derived from general protected area assessments, such as that of Bruner et al. (2004). It needs to be calculated, as in this analysis, based on what is actually required at specified sites to protect and monitor CHV species (Bruner et al., 2004). Proxy measures, remote sensing, or modeling is not helpful (Walston et al., 2010). Walston et al. (2010) estimated that the total amount of money required for protecting and monitoring tigers and their prey was US$82 million per year (or ~US$930 per km^2 per year). However, they also show that more than half of that sum (US$47 million per year) was already being committed by range state governments and, to a far lesser extent, international donors and NGOs, leaving an overall shortfall of US$35 million per year for all tiger source sites.

As we show in this paper, significant overlap in the ranges of CHV species in protected areas reduces the protection and monitoring costs per CHV species: the increased cost per square kilometer per year when one or more of the six nontiger CHV species are present in addition to tigers is modest (1.0–1.15 the cost of just protecting tigers and their prey), giving a total of US$89 million per year for conservation of tigers and the other CHV species in the 42 tiger source sites and a grand total of approximately US$387 million per year (equivalent to US$954 per km^2 per year) for conservation of all seven CHV species (including tigers) in all the protected areas in which they occur.

We were not able to repeat the compilation and verification of government and other budgets conducted by Walston et al. (2010) for those protected areas not included in their analysis (i.e., we do not have budgets for protected areas without tigers) or for the few species-specific activities relevant to CHV species other than tigers. For protected areas with tigers, this is likely to result in an insignificant underestimation of available budgets since with the exception of HEC mitigation work, there probably is not a great deal of money committed beyond what is already included in the analysis for tigers (e.g., infrastructure, staff, law enforcement and monitoring). Thus, the overall shortfall is probably not a great deal more than the US$35 million estimated by Walston et al. (2010).

For protected areas without tigers, the underestimate is significant, but its effect will be to overestimate the funding needed for CHV conservation because significantly more money has actually been committed by governments and other agencies for CHV species conservation in protected areas across Asia than the US$47 million reported by Walston et al. (2010) for tiger source sites. For example, Ahmad Zafir et al. (2011) report that some US$1.25 million per year is spent on Sumatran rhinoceros conservation in four priority protected areas.

Investments in CHV species conservation should be focused on protected areas

Since few protected areas in Asia currently meet all the criteria for the effective conservation of CHV species, raising standards in existing protected areas is the highest priority. There is a potential for achieving very significant increases in tiger, rhinoceros, elephant, gaur, and banteng numbers through relatively modest investments in and improvements to existing protected areas across Asia. Creating new protected areas is both more expensive and much more difficult from a sociopolitical perspective than improving existing protected areas. Perhaps most critically, without effective management, simple increases in the number of protected areas provide negligible benefits (Collen et al. this volume, Chapter 6). New protected areas should not, therefore, be created at the expense of improving existing protected areas for CHV species. However, where CHV species occur outside of protected areas and the criteria for effective management we present in this paper can be met, establishing new protected areas should be considered in order to inter alia help achieve representation, resiliency, and redundancy (Shaffer & Stein, 2000) and to facilitate metapopulation management (Thapa et al., 2013).

The "landscape approach" versus the "protected area" (and "source site") approach to CHV species conservation: A false dichotomy?

Protecting CHV species in protected areas is essential to reverse the declines in these species but is just one element of a long-term conservation strategy. For large, wide-ranging species, such as tigers and elephants, conservation planning at the landscape level is also necessary. Many protected areas in Asia are small, so the conservation community should (i) seek to identify the largest areas where protection can be provided, at the moment that almost always means protected areas, although there are a few encouraging examples of community-based law enforcement in wildlife corridors in Nepal (Thapa et al., 2013), and (ii) endeavor to ensure that the land use regimes around protected areas are as "wildlife friendly" as possible. One promising approach is that of the "Managed Elephant Range" (MER); the MER concept provides a landscape-level approach in which planners assess the habitat requirements of elephants over large areas and allow for compatible human activities such as reduced impact forestry, slow rotation shifting cultivation, and controlled livestock grazing in some zones. MERs are typically established outside of, usually as extensions to, existing protected areas, which are treated as core areas providing for strict protection embedded in the larger MER (Olivier, 1978; McNeely & Sinha, 1981). Malaysia, for example, has adopted the MER concept, both in Peninsular Malaysia and in Sabah (SWD, 2011; Samsudin et al., in review). The "tiger-permeable landscapes" of Walston et al. (2010) represent essentially the same concept as MERs, but for tigers. In Nepal and India, there are now at least two metapopulations of the greater one-horned rhinoceros in the western part of the "Terai." One is based around Bardia National Park and Katerniaghat Wildlife Sanctuary, which serve as core areas, with rhinoceroses using the Khata Corridor linking the two protected areas; the other is around the cluster of protected areas consisting of Suklaphanta Wildlife Reserve, Dudhwa National Park, Lagga Bagga Forest Reserve, and Kishanpur Wildlife Sanctuary (Thapa et al., 2013). Core areas with strict protection are, nevertheless, vital stepping stones to all such large landscape-based initiatives.

Conclusions

The seven Asian CHV species discussed in this chapter share a specific, very significant threat: high levels of illegal hunting as a result of their commercial value in international markets. As a result, these species are now largely or wholly confined to protected areas with substantial and almost certainly increasing overlap in their range.

While the loss of habitat is a major challenge for these and other species in Asia, large swathes of suitable habitat are now bereft of CHV species due to illegal hunting. The tiger presents the clearest example of a species for which considerable habitat remains within its historical range, but which is now mostly absent from 93% of these areas, due to illegal hunting, both of tigers and their prey. Further particularly illustrative cases include rhinoceroses across the Southeast Asian mainland and Borneo, banteng outside of Java, and elephants in Myanmar.

Even within protected areas, the prevention of habitat loss has been more successful than the prevention of illegal hunting. The need for effective law enforcement focused specifically on CHV species is essential and lacking in many places. The importance of protected areas and law enforcement therein is only going to increase as the demand for CHV species and their parts grows, and demand seems unlikely to decline significantly in the immediately foreseeable future.

Unfortunately, few protected areas in Asia currently meet the criteria for effective conservation of CHV species, so raising standards in existing protected areas is the highest priority. The conservation community should also be seeking to expand the protected area network in Asia (in the interests of achieving representation, resiliency, and redundancy and to help facilitate metapopulation management) but not at the expense of effective protection of existing protected areas where these species occur.

Finally, effective protection of CHV species is more expensive than is often recognized, but the costs are attainable and range state governments are already contributing much more than are acknowledged (often more than 50% of the budgets required). This provides a sound foundation for external investments. The overlapping distribution of CHV species in Asian protected areas and the modest additional species-specific funding needs demonstrated in this chapter shows that a relatively small increase in existing investments to protected areas for protection would deliver high conservation returns.

Acknowledgments

We would like to thank the organizers of the November 2012 "Protected Areas: Are They Safeguarding Biodiversity?" Symposium at the Zoological Society of London: Lucas Joppa (Microsoft Research), John Robinson (Wildlife Conservation Society), and Jonathan Baillie (Zoological Society of London). We would also like to thank the IUCN/SSC Asian Elephant Specialist Group and the IUCN/SSC Asian Wild Cattle Specialist Group for providing data and the Global Environment Facility (GEF), the US Fish and Wildlife Service, and the Wildlife Conservation Society. We acknowledge all the contributors to the Walston et al. (2010) analysis. Finally, we thank Hannah Stevens for providing GIS support for some of our early analyses.

Note

1. The tiger "source sites" of Walston et al. (2010) represent a subset of protected areas containing tigers that were selected primarily for the importance of their tiger populations; source sites represent 42 protected areas with tigers and account for almost 70% of the global tiger population (please see Walston et al., 2010 for further detail).

References

Ahmad Zafir, A. W. et al., 2011. Now or never: What will it take to save the Sumatran rhinoceros (*Dicerorhinus sumatrensis*) from extinction. *Oryx*, Volume 45, p. 225–233.

Bindra, P. & Karanth, K. K., 2013. Tourism turf wars: Debating the benefits and costs of wildlife tourism in India. *Oryx*, Volume 47, p. 15–19.

Blake, S. & Hedges, S., 2004. Sinking the flagship: The case of forest elephants in Asia and Africa. *Conservation Biology*, Volume 18, p. 1191–1202.

Brook, S. M. et al., 2012. Integrated and novel survey methods for rhinoceros populations confirm the extinction of *Rhinoceros sondaicus annamiticus* from Vietnam. *Biological Conservation*, Volume 155, p. 59–67.

Bruner, A., Gullison, R., & Balmford, A., 2004. Financial costs and shortfalls of managing and expanding protected area systems in developing countries. *BioScience*, Volume 54, p. 1119–1126.

CITES, 2013. Monitoring the illegal killing of elephants: A report to the standing committee of CITES 16th meeting (Bangkok, March 2013). CoP 16 Doc.53.1, Geneva: CITES.

Coad, L. et al., 2013. Progression towards the CBD protected area management effectiveness targets. *PARKS*, Volume 19(1), p. 13–24.

Dinerstein, E., 2003. *The return of the unicorns: The natural history and conservation of the greater one-horned rhinoceros*, New York: Columbia University Press.

DNPWC, 2000. *Rhino count 2000: Initial report*, Kathmandu: DNPWC.

Duckworth, J. W. & Hedges, S., 1998. *Tracking tigers: A review of the status of tiger, Asian elephant, Gaur and Banteng in Vietnam, Lao, Cambodia and Yunnan (China) with recommendations for future conservation action*, Hanoi: WWF.

Duckworth, J. W. et al., 2008. Bos gaurus: IUCN Red List of Threatened Species 2011. [Online] Available at: http://www.iucnredlist.org [Accessed December 12, 2011].

Dudley, N., 2008. *Guidelines for applying protected area management categories*, Gland: IUCN.

Dudley, N., Stolton, S., & Elliot, W., 2013. Wildlife crime poses unique challengers to protected areas. *PARKS*, Volume 19, p. 7–12.

Emslie, R. H., Milliken, T., & Talukdar, B. K., 2013. African and Asian rhinoceroses: Status, conservation and trade. Report to CITES 16th meeting (Bangkok, March 2013). CoP Doc.54.2, Geneva: CITES.

ESRI, 2011. *ArcGIS desktop: Release 10*, Redlands: Environmental Systems Research Institute.

Goossens, B. et al., 2013. Genetics and the last stand of the Sumatran rhinoceros (*Dicerorhinus sumatrensis*). *Oryx*, Volume 47, p. 340–344.

Gopala, A. et al., 2011. Elephas maximus ssp. sumatranus: IUCN Red List of Threatened Species 2011. [Online] Available at: http://www.iucnredlist.org [Accessed December 6, 2011].

Gunaryadi, D., Sugiyo, & Hedges, S., in review. Community based human elephant conflict mitigation: The value of an evidence based approach in promoting the uptake of effective methods. *PLoS ONE*.

Haken, J., 2011. *Transnational crime in the developing world*, Washington, DC: Global Financial Integrity.

Hedges, S., 1995. *Banteng (*Bos javanicus*): Species account and recommendations for conservation*, Gland: IUCN SSC Asian Wild Cattle Specialist Group.

Hedges, S., 2006. Conservation. In: M. E. Fowler & S. K. Mikota, eds. *Biology, medicine and surgery of elephants*. Ames: Blackwell Publishing, p. 475–489.

Hedges, S., 2012. *Monitoring elephants and assessing threats: A manual for researchers, managers and conservationists*, Himayatnagar: Universities Press.

Hedges, S. & Gunaryadi, D., 2010. Reducing human-elephant conflict: Do chillies help deter elephants from entering crop fields. *Oryx*, Volume 44, p. 139–146.

Hedges, S., Tyson, M. J., Sitompul, A. F., & Hammatt, H., 2006. Why inter country loads will not help Sumatra's elephants. *Zoo Biology*, Volume 25, p. 235–246.

Hedges, S., Fisher, K., & Rose, R., 2009. Range wide priority setting workshop for Asian elephants (*Elephas maximus*). Final report: December 2009. A report to the US Fish & Wildlife Service, Bronx: WCS.

Hedges, S. et al., 2013. Accuracy, precision and cost effectiveness of conventional dung density and faecal DNA based survey methods to estimate Asian elephant (*Elephas maximus*) population size and structure. *Biological Conservation*, Volume 159, p. 101–108.

Hoogerwerf, A., 1979. *Udjung Kulon: The land of the last Javan rhinoceros*, Leiden: EJ Brill.

Hubschle, A., 2014. Of bogus hunters, queenpins and mules: The varied roles of women in transnational organized crime in southern Africa. *Trends in Organized Crime*, Volume 17(1–2, p. 31–51.

IUCN, 2010. *Regional conservation strategy for wild cattle and buffaloes in Southeast Asia*, Gland & Cambridge: IUCN SSC AWCSG.

Karanth, K. U., 2002. Nagarahole: Limits and opportunities in wildlife conservation. In: J. Terborgh, C. v. Schaik, L. Davenport, & M. Rao, eds. *Making parks work: Strategies for preserving tropical nature*. Washington, DC: Island Press, p. 189–202.

Karanth, K. U. & Nichols, J. D., 2010. Non invasive survey methods for assessing tiger populations. In: R. Tilson & P. Nyhus, eds. *Tigers of the world: The science, politics and conservation of Panthera tigris*. London: Academic Press, p. 241–261.

Karanth, K. U. et al., 2003. Science deficiency in conservation practice: The monitoring of tiger populations in India. *Animal Conservation*, Volume 6, p. 141–146.

Karanth, K. U., Nichols, J. D., Kumar, N. S., & Hines, J. E., 2006. Assessing tiger population dynamics using photographic capture-recapture sampling. *Ecology*, Volume 87, p. 2925–2935.

Leimgruber, P. et al., 2003. Fragmentation of Asia's remaining wildlands: Implications for Asian elephant conservation. *Animal Conservation*, Volume 6, p. 347–359.

Leimgruber, P. et al., 2008. Modeling population viability of captive elephants in Myanmar (Burma): Implications for wild populations. *Animal Conservation*, Volume 11, p. 198–205.

Martin, E. B., 2010. *From the jungle to Kathmandu: Horn and tusk trade*, Lalitpur: WWG.

Martin, E. & Vigne, L., 2011. *The ivory dynasty: A report on the soaring demand for elephant and mammoth ivory in southern China*, London: Elephant Family, The Aspinall Foundation & Columbus Zoo and Aquarium.

Mason, C. F., Bulte, E. H., & Horan, R. D., 2012. Banking on extinction: Endangered species and speculation. *Oxford Review of Economic Policy*, Volume 28, p. 180–192.

McNeely, J. A. & Sinha, M. K., 1981. Protected areas for Asian elephants. *PARKS*, Volume 6, p. 4–7.

McWilliam, A., Hedges, S., Johnson, A., & Luangyotha, P., 2010. *A manual for human-elephant conflict monitoring*, Vientiane: WCS.

Milliken, T. & Shaw, J., 2012. *The South Africa-Vietnam rhino horn trade nexus: A deadly combination of institutional lapses, corrupt wildlife industry professionals and Asian crime syndicates*, Johannesburg: TRAFFIC.

Milliken, T., Emslie, R. H., & Talukdar, B. K., 2009. African and Asian rhinoceroses: Status, conservation and trade. Report to CITES 15th meeting (Doha, March 2010). CoP 15 Doc.45.1a, Geneva: CITES.

Mills, J. A. & Jackson, P., 1994. *Killed for a cure: A review of the worldwide trade in tiger bone*, Cambridge: TRAFFIC.

Miquelle, D. G. et al., 2005. Searching for the coexistence recipe: A case study of conflicts between people and tigers in the Russian Far East. In: R. Woodroffe, S. Thirgood, & A. Rabinowitz, eds. *People and wildlife: Conflict or coexistence*. Cambridge: Cambridge University Press, p. 305–322.

Moyle, B., 2009. The black market in China for tiger products. *Global Crime*, Volume 10, p. 124–143.

O'Kelly, H. J. et al., 2012. Identifying conservation successes, failures and future opportunities: Assessing recovery potential of wild ungulates and tigers in eastern Cambodia. *PLoS ONE*, Volume 7, p. e40482.

Olivier, R. C. D., 1978. Distribution and status of the Asian elephant. *Oryx*, Volume 14, p. 379–424.

Pedrono, M., Tuan, H. M., Chouteau, P., & Vallejo, F., 2009. Status and distribution of the endangered banteng (*Bos javanicus birmanicus*) in Vietnam: A conservation tragedy. *Oryx*, Volume 43, p. 618–625.

Rademeyer, J., 2012. *Killing for profit: Exposing the illegal rhino horn trade*, Cape Town: Zebra Press.

Rosen, G. & Smith, K., 2010. Summarizing the evidence on the international trade in illegal wildlife. *EcoHealth*, Volume 7, p. 24–32.

Samsudin, A. R. et al., in review. Viability and management of the Asian elephant populations in the Endau Rompin landscape, Malaysia. *PLoS ONE*.

Sanderson, E. W. et al., 2006. *Setting priorities for the conservation and recovery of wild tigers: 2005–2015: The technical assessment*, Washington, DC: Save the Tiger Fund, WCS & WWF.

't Sas-Rolfes, M. & Fitzgerald, T., 2013. *Can a legal horn trade save rhinos*, Bozeman: PERC.

Seidensticker, J., Jackson, P., & Christie, S., 1999. *Riding the tiger: Tiger conservation in human dominated landscapes*, Cambridge: Cambridge University Press.

Shaffer, M. L. & Stein, B. A., 2000. Safeguarding our precious heritage. In: B. A. Stein, L. S. Kutner, & J. S. Adams, eds. *Precious heritage: The status of biodiversity in the United States*. New York: Oxford University Press, p. 301–322.

Shepherd, C. R. & Nijman, V., 2008. *Elephant and ivory trade in Myanmar*, Petaling Jaya: TRAFFIC.

Srikosamatara, S. & Suteethorn, V., 1994. Wildlife conservation along the Thai-Lao Border. *Natural History Bulletin of the Siam Society*, Volume 42, p. 321.

Srikosamatara, S., Siripholdej, B., & Suteethorn, V., 1992. Wildlife trade in Lao PDR and between Lao PDR and Thailand. *Natural History Bulletin of the Siam Society*, Volume 40, p. 1–47.

Stiles, D., 2009. *The elephant and ivory trade in Thailand*, Petaling Jaya: TRAFFIC.

Subedi, N. et al., 2013. Population status, structure and distribution of the greater one-horned rhinoceros. *Rhinoceros unicornis* in Nepal. *Oryx*, Volume 47, p. 352–360.

Sukumar, R., 2003. *The living elephants: Evolutionary ecology, behaviour and conservation*, Oxford: Oxford University Press.

SWD, 2011. *Elephant action plan: 2012–2016*, Kota Kinabalu: Sabah Wildlife Department.

Talukdar, B. K., Sectionov, & Whetham, L. B., 2010. Proceedings of the Asian Rhino Specialist Group meeting held at Kaziranga National Park (India, February 2010), Gland: IUCN SSC AsRSG.

Thapa, K. et al., 2013. Past, present and future conservation of the greater one-horned rhinoceros (*Rhinoceros unicornis*) in Nepal. *Oryx*, Volume 47, p. 345–351.

Tilson, R. & Nyhus, P. J., 2010. *Tigers of the World: The science, politics and conservation of Panthera tigris*, London: Academic Press.

Van Song, N., 2008. Wildlife trading in Vietnam: Situation, causes and solutions. *The Journal of Environment and Development*, Volume 17, p. 145–165.

Vandegrift, J., 2013. Elephant poaching: CITES failure to combat the growth in Chinese demand for ivory. *Virginia Environmental Law Journal*, Volume 31, p. 102–136.

Walston, J. et al., 2010. Bringing the tiger back from the brink: The six percent solution. *PLoS Biology*, 8(9), p. e1000485.

WDPA, 2010. *The world database on protected areas (WDPA)*, Gland & Cambridge: IUCN & UNEP-WCMC.

Part IV
Monitoring Protected Areas at System Scales

12

Monitoring Protected Area Coverage and Impact on Key Biodiversity Areas, Important Bird Areas and Alliance for Zero Extinction Sites

S. H. M. Butchart[1], *T. M. Brooks*[2], *J. P. W. Scharlemann*[3,4] *and M. A. K. Mwangi*[1]

[1]BirdLife International, Cambridge, UK
[2]NatureServe, Arlington, VA, USA
IUCN, Gland, Switzerland
[3]United Nations Environment Programme World Conservation Monitoring Centre, Cambridge, UK
[4]School of Life Sciences, University of Sussex, Brighton, UK

Introduction

Through the Convention on Biological Diversity (CBD)'s Strategic Plan, the world's governments have committed to conserve effectively 17% of terrestrial (and inland water) areas by 2020, 'especially areas of particular importance for biodiversity' (CBD, 2010). In this chapter, we consider how governments can identify such areas in order to focus their efforts in expanding protected area networks and how they can monitor the effectiveness of protected areas that overlap such sites.

While a number of broad-scale schemes addressing spatial priority setting for biodiversity conservation have been published in recent decades, ranging from hotspots to wilderness areas and ecoregions (Brooks et al., 2006), they are not particularly useful for pinpointing the most important sites for conserving biodiversity, including through protected areas. In response, several global or regional initiatives have emerged to identify, using systematic methods, networks of sites that are of particular significance for the persistence of global biodiversity. These

Protected Areas: Are They Safeguarding Biodiversity?, First Edition.
Edited by Lucas N. Joppa, Jonathan E. M. Baillie and John G. Robinson.

were pioneered by BirdLife International for birds, Important Bird and Biodiversity Areas (IBAs) (BirdLife International, 2010), but have also been developed for plants, Important Plant Areas (IPAs) (PlantLife International, 2004), freshwater taxa (Holland et al., 2012) European butterflies (Van Swaay & Warren, 2003) and highly threatened taxa restricted to single sites, Alliance for Zero Extinction sites (AZEs) (Ricketts et al., 2005). These initiatives are being brought together under the umbrella concept of Key Biodiversity Areas (Eken et al., 2004; Langhammer et al., 2007), with a joint IUCN Species Survival Commission (IUCN SSC) and World Commission on Protected Areas (WCPA) task force currently developing standardised criteria and guidance.

At present, only two of these networks of important sites for biodiversity have been identified worldwide, IBAs and AZEs. IBAs are places of international significance for the conservation of birds; 10,993 such sites have been identified based on their populations of one or more of 4,445 threatened, restricted range, biome-restricted or congregatory species. IBAs are identified, usually at a national scale through multi-stakeholder processes, using a standardised set of data-driven criteria and thresholds, relating to threatened, restricted range, biome-restricted and congregatory species. IBAs are delimited so that, as far as possible, they:

- Are different in character, habitat or ornithological importance from surrounding areas
- Provide the requirements of the trigger species (i.e. those for which the site qualifies) while present, alone or in combination with networks of other sites
- Are or can be managed in some way for conservation (Fishpool et al., 1998; BirdLife International, 2010)

IBAs have been identified in almost all countries of the world (Evans, 1994; Evans & Heath, 2000; Fishpool & Evans, 2001; Chan et al., 2004; Davenish et al., 2009; Butchart et al., 2012).

AZEs are sites meeting three criteria:

1. *Endangerment*: Supporting at least one endangered or critically endangered species, as listed on the IUCN Red List
2. *Irreplaceability*: Holding the sole or overwhelmingly significant (≥95%) known population of the target species, for at least one life history segment
3. *Discreteness*: Having a definable boundary within which the character of habitats, biological communities and or management issues have more in common with each other than they do with those in adjacent areas (Ricketts et al., 2005)

Hence, AZEs represent locations at which species extinctions are imminent unless appropriately safeguarded (i.e. protected or managed sustainably in ways consistent with the persistence of populations of target species). A total of 588 terrestrial AZE sites have been identified globally for 919 highly threatened mammals, birds, amphibians, reptiles from selected clades (Testudines, Crocodylia and Iguanidae) and conifers (Butchart et al., 2012; American Bird Conservancy, 2013).

The IBA and AZE site networks are, by definition, areas of particular importance for biodiversity as referred to in the CBD target. Hence, they represent priority areas to consider for designation as formal protected areas or other site safeguard mechanisms.

Here, we describe recent trends and the current situation in terms of coverage of these two site networks by protected areas and the impact of this protection on their status. Specifically, we quantify:

- The degree of coverage by protected areas of IBAs and AZEs and for those sites that are protected
- Trends in the state, pressure and conservation responses
- Trends in extinction risk of species for which the sites were identified

In doing so, we draw heavily on Mwangi et al. (2010) and Butchart et al. (2012), to which we refer the reader for further details, including methods underlying the analyses described (Mwangi et al., 2010; Butchart et al., 2012).

Protected area coverage of important sites for biodiversity conservation

Butchart et al. (2012) showed that by 2010, only 28% of IBAs were completely covered by protected areas and 49% were wholly unprotected; on average, only 39% of the area of each IBA was protected (Figure 12.1). Protected area coverage of IBAs was lowest in the Brazilian Atlantic Forest, Middle East, Northern Africa, freshwater ecosystems and deserts. AZEs were marginally less well covered by protected areas; 22% of sites were completely covered, 51% were unprotected and 35% of the areas of each site were covered on average (Figure 12.1).

Nevertheless, the current figures represent a substantial growth in coverage in recent years. For example, the proportion of IBAs completely protected is estimated to have been 21% in 1990 and 27% in 2000, with a mean area protected of 28 and 36%, respectively, for these dates. For AZEs, the proportion of sites completely protected is estimated to have been 17% in 1990 and 21% in 2000, with a mean area protected of 28 and 33%, respectively, for these dates.

While the coverage of IBAs and AZEs has grown, there has been a concomitant growth in protected area coverage in general, and the proportion of the total protected area extent that covers important sites has declined significantly since 1950 for both IBAs (annual change between −0.45 and −1.14%, $P<0.001$, $N=57$ years) and AZEs (−0.79 and −1.49%, $P<0.001$, $N=57$ years) (Figure 12.2).

Protected areas are therefore increasingly being designated outside important sites for species conservation, despite the high proportion of such sites that have

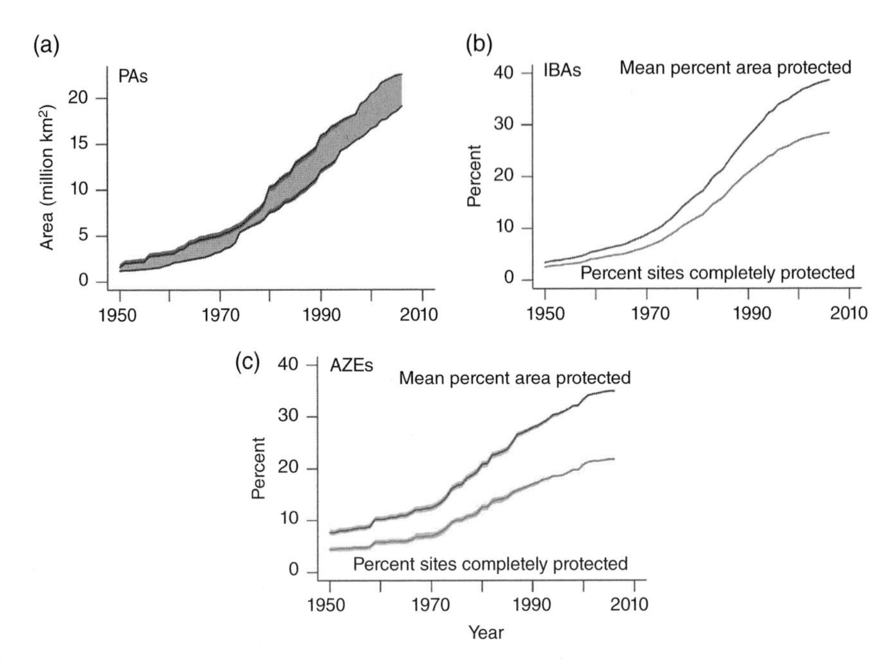

Figure 12.1 **Trends in (a) the extent of terrestrial protected areas and their coverage of (b) IBAs and (c) AZEs. For protected areas, the lines represent minimum and maximum estimates with 95% confidence intervals, derived from areas with delimited boundaries and those with and without delimited boundaries, respectively. For IBAs and AZEs, shading shows 95% confidence intervals based on uncertainty around date of protection (and, for a small subset of IBAs, proportion protected). Source: Reproduced with permission from Butchart et al. (2012)**

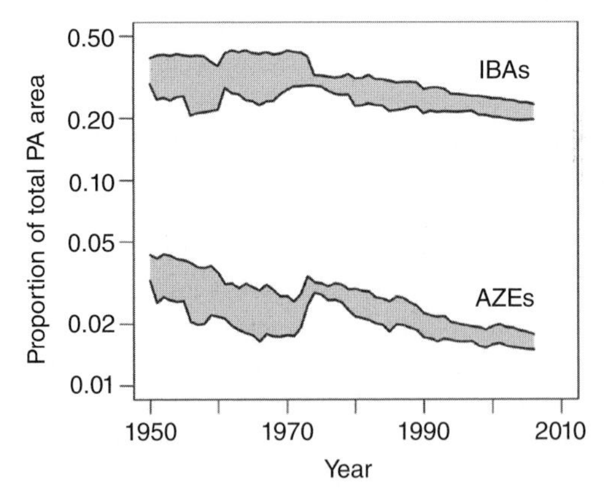

Figure 12.2 **The proportion of total protected area extent covering important sites, 1950–2006. Lines represent minimum and maximum estimates based on uncertainty in the extent of protected areas. Source: Reproduced with permission from Butchart et al. (2012)**

yet to be protected. This may have occurred due to a number of interrelated reasons. One explanation based on previous literature in this field is that protected areas tend to be biased to higher elevations, steeper slopes, greater remoteness and areas with lower suitability for agriculture (Pressey et al., 2002; Joppa & Pfaff, 2009), which are not necessarily the areas of greatest importance for biodiversity, nor the ones where rapid or extensive land use change is occurring. Areas of highest biological value are often more financially, socially and politically costly to protect, because of covariance between species richness, endemism and human population density (Luck, 2007). However, we found that the expected negative correlation between the proportion of IBAs protected and local human population density was rather weak, although stronger in developing countries (Butchart et al., 2012). Yet this explanation may become increasingly important in future, as sites of high economic and biodiversity value are increasingly converted to agriculture and other extractive uses.

That the increased coverage of protected areas has not had a proportional impact on IBAs and AZEs might indicate that some governments lack awareness of, or are reticent to use, information on IBAs and AZEs in protected area planning. The identification and documentation of IBAs and AZEs occurred relatively recently (since the 1980s for IBAs and 2004 for AZEs), although the conservation importance and need for protection of many of these sites have been known for decades (Butchart et al., 2012). However, in some countries, IBA inventories have played an important role in informing recent protected area designation (e.g. Madagascar, Philippines and the European Union) (BirdLife International, 2010) or expansion (e.g. Nicaragua) (Davenish et al., 2009). Finally, some protected areas may have been targeted primarily at wilderness areas, abiotic processes (e.g. hydrological) or locations for recreation, tourism, hunting, scenery or cultural interest rather than biodiversity itself (Joppa & Pfaff, 2009). Exploring which of these explanations are the most important, and which barriers to site protection are most significant, would be useful avenues for future research.

State, pressure and responses at protected areas that overlap important sites for biodiversity conservation

In recent years, BirdLife International developed, trialled and implemented a programme of monitoring within IBAs (Bennun et al., 2005). The monitoring is designed to be simple, sustainable and cost-effective to maximise local participation and institutional engagement. A simple method is used to assign scores (0–3) for the condition of IBAs (state), the threats they face (pressure) and the conservation action being taken (response). State is scored based on population trends of species for which the site was identified, or habitat condition as a surrogate. Pressure is categorised following Salafsky et al. (2008) and scored based on timing, scope (the proportion of the population of trigger species at the site affected) and severity (the rate of decline caused by the threat) (Salafsky et al., 2008). Response is scored in relation to the level of formal protection, management planning and implementation of action (BirdLife International, 2006). This monitoring framework of state–pressure–response is a simplified version of that used by the CBD's Strategic Plan (Sparks et al., 2011).

Mwangi et al. (2010) provided the first published example of indicators derived from IBA monitoring, using Kenya as a case study. Here, we discuss their results to demonstrate how this approach can be applied to protected areas and to compare recent trends in state, pressure and response for protected versus unprotected IBAs during 1999–2005.

Mwangi et al. (2010) derived data for 36 of the national total of 60 IBAs from Bennun & Njoroge (1999), Ng'weno et al. (2004) and Musila et al. (2006), with these sources presenting results from data collected by national and local institutions with field staff and community groups at IBAs (Bennun & Njoroge, 1999; Ng'weno et al., 2004; Musila et al., 2006). As the global methodology for IBA monitoring evolved during the period 1999–2005, the published scores could not be used directly to calculate IBA indices, so the information presented was used to determine retrospectively standardised scores.

For IBAs with more than one trigger species occupying different types of habitats, state scores were derived for the habitat of each trigger species. To determine the overall score for the IBA, a 'weakest link' approach was used; that is, the lowest (worst) condition for any key habitat for a trigger species determined the overall score for the IBA. Similarly, the highest impact threat to any trigger species determined the overall pressure score for the IBA. This simple decision rule follows the precautionary approach adopted by the scoring methodology.

Overall, there was a decline in state and pressure scores for Kenyan IBAs from 1999 to 2005, although this was not significant (Kruskal–Wallis test: state $H=2.46$, pressure $H=0.46$, all $P>0.05$, $n=36$). However, response scores increased significantly from 1999 to 2005 ($H=13.55$, all $P=0.001$, $n=36$). A comparison of trends for protected versus unprotected sites showed that scores for the state of protected IBAs improved slightly during 1999–2005, compared to a negative trend in the state of unprotected IBAs (Figure 12.3). By 2005, protected IBAs ($n=20$) were in better condition than those that were unprotected ($n=16$), though this difference was not statistically significant (Mann–Whitney test: $U=0.111$, $P=0.11$). As might be expected, protected IBAs had significantly higher scores for response than unprotected IBAs ($U=24$, $P=0.001$) (Figure 12.3). During 1999–2005, there was a 55% increase in the number of protected areas that had comprehensive and up-to-date management plans and 25% increase in comprehensive conservation action at these sites. By 2005, only one unprotected IBA had a comprehensive management plan, but it still lacked comprehensive conservation action. However, limited substantial conservation activities were recorded in 44% of the unprotected IBAs by 2005, representing a 38% increase from 1999 (Figures 12.4 and 12.5). Despite the differences in responses, there was no significant difference in

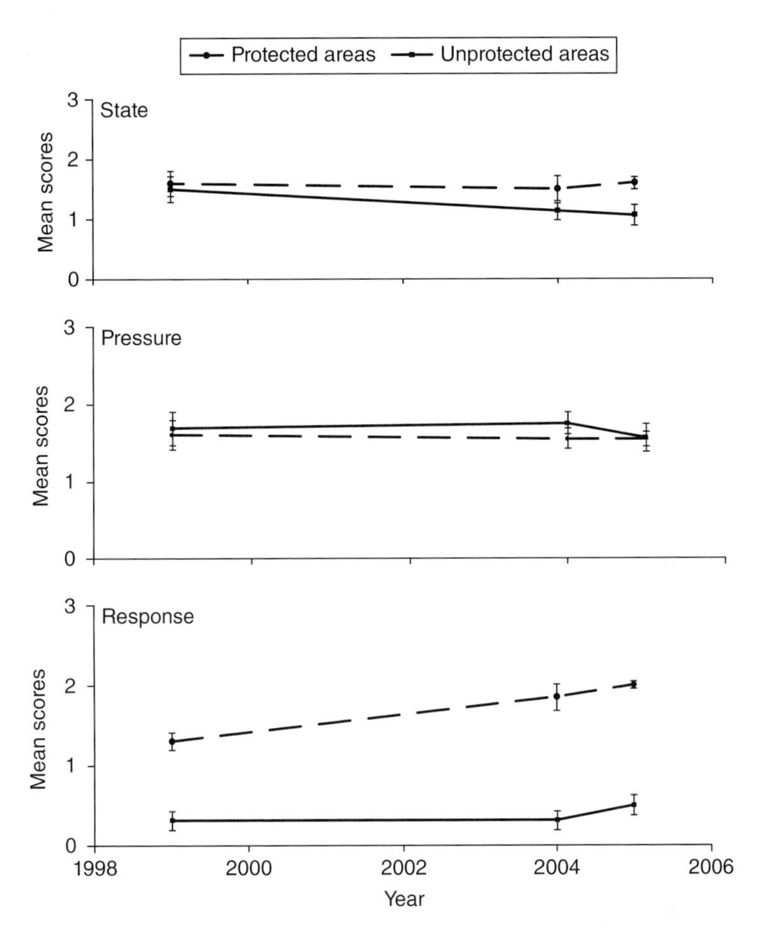

Figure 12.3 **IBA indices (mean ± 95% CI) for state, pressure and response during 1999–2005 at protected (filled circles, $n = 20$) and unprotected (filled squares, $n = 16$) Kenyan IBAs. Source: Adapted from Mwangi et al. (2010). Reprinted with permission of Cambridge University Press**

scores for pressure at protected versus unprotected IBAs ($U = 160$, $P > 0.05$) (Figure 12.3).

The overall trends in the state–pressure–response at Kenyan IBAs during 1999–2005, as measured using the BirdLife monitoring framework, are consistent with the changes in the policy and management of natural resources that occurred in Kenya during this period. The national population grew by 2.9% during 1989–1999 (KNBS, 2010), increasing demand often to beyond sustainable levels for natural resources from the habitats in which IBAs are located (NES, 2000). As a result of the growing pressures, a suite of changes in policy and the management of natural resources was commissioned by the Kenyan government

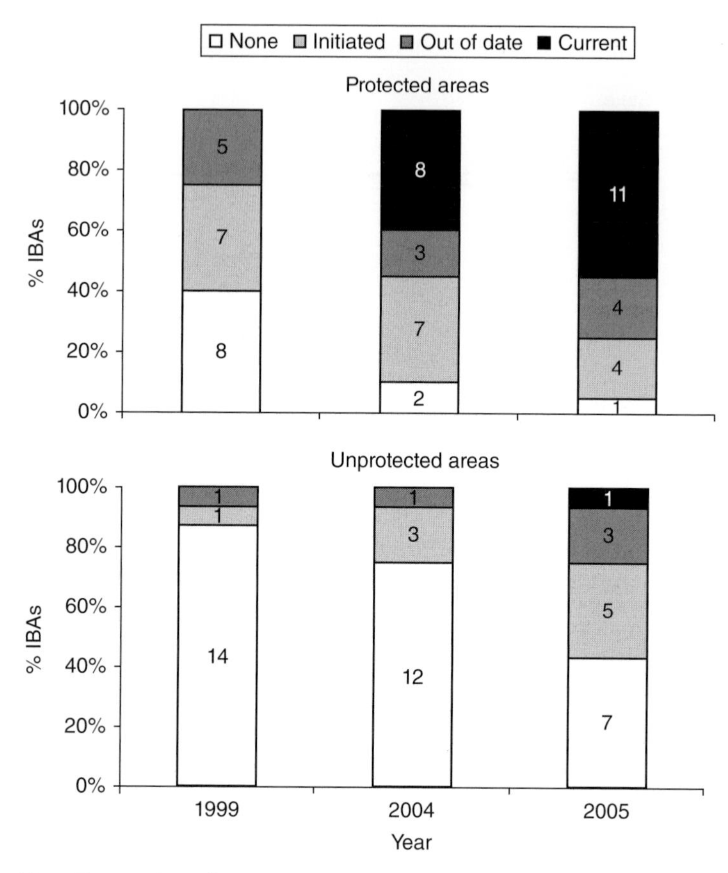

Figure 12.4 **Proportion of Kenyan IBAs that are protected areas ($n=20$) or unprotected ($n=16$) with different categories of scores for management planning during 1999–2005; figures give the number of IBAs. Source: Adapted from Mwangi et al. (2010). Reprinted with permission of Cambridge University Press.**

during the period considered here, including the establishment of the National Environment Management Authority and Public Complaints Committee, publication of Environmental Impact Assessment Guidelines, initiation of the Wetlands Conservation and Management Policy, review of the Geology and Mining Act and passing of the Forest Act. These appear to have contributed to the improvement in IBA management (especially in protected areas) as shown by the mean response score.

Protected Kenyan IBAs had substantially greater conservation responses and consequently appear to be in marginally better condition than unprotected IBAs (with lower pressures, although not significantly so). This supports arguments for the government to continue to invest resources in current protected areas, expand

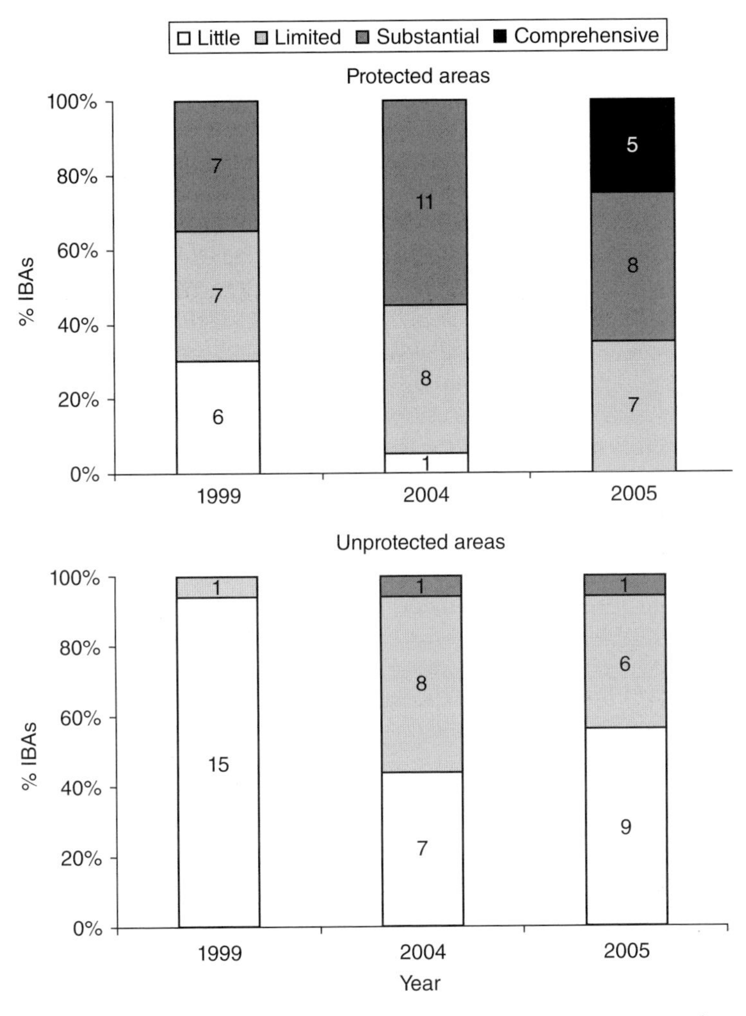

Figure 12.5 **Proportion of Kenyan IBAs that are protected areas IBAs ($n = 20$) or unprotected ($n = 16$) with different categories of scores for conservation action during 1999–2005; figures give the number of IBAs. Source: Adapted from Mwangi et al. (2010). Reprinted with permission of Cambridge University Press**

the protected area network and work with local communities to conserve biodiversity. There was also some evidence that protected areas managed jointly by more than one agency may have fared better than those managed by one alone; the state of protected IBAs managed jointly by more than one agency significantly improved, while that of IBAs managed individually by the Kenya Forest Service, Kenya Wildlife Service or National Museums of Kenya declined or remained

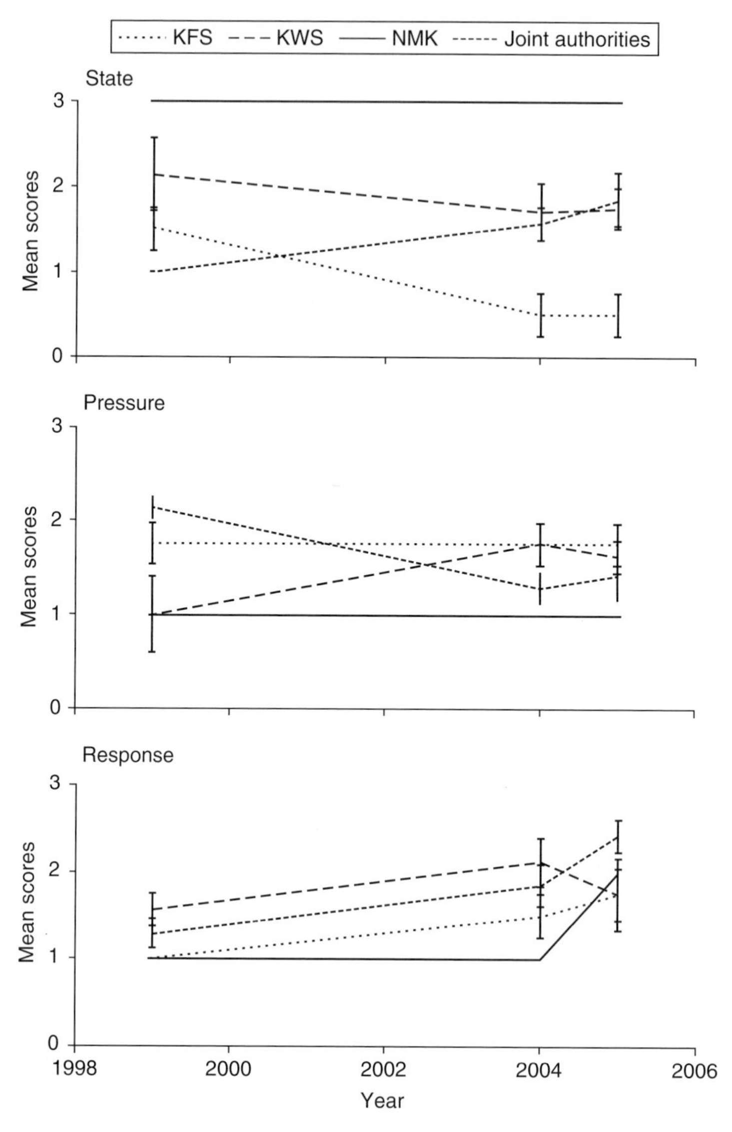

Figure 12.6 **IBA indices (mean ± 95% CI) for state, pressure and response during 1999–2005 at protected Kenyan IBAs managed by the Kenya Forest Service (KFS; $n = 4$), Kenya Wildlife Service (KWS; $n = 7$), National Museums of Kenya (NMK; $n = 1$) or jointly by more than one of these agencies (JA; $n = 8$). Source: Adapted from Mwangi et al. (2010). Reprinted with permission of Cambridge University Press.**

stable (Figure 12.6). Jointly managed sites showed improvements in responses, a greater reduction in pressures and the largest improvements in state, compared to sites managed by single agencies, presumably because of the benefits of pooling resources to fund management of these protected areas.

These results for Kenya also have wider implications for protected area monitoring. Limited access to data and information and lack of adequate infrastructure to enhance their collection have been identified as hindrances to better utilisation and management of biodiversity in Kenya (NES, 1994, 2000), a finding that applies across much of the developing world. The BirdLife IBA monitoring system is sensitive enough to detect differences between sites and over time but is sufficiently simple to be implemented with limited training and resources and without sophisticated technology. The results provide useful information for managers of individual protected areas, management agencies responsible for suites of sites and national governments. IBA monitoring has great potential to be used by governments to assess management effectiveness of their protected area networks.

Trends in the extinction risk of species in protected areas that overlap important sites for biodiversity conservation

A second approach to assessing the impact of protected areas on biodiversity is to use the Red List Index (Butchart et al., 2004, 2005, 2007, 2012). This indicator is based on the number of species in each Red List Category of extinction risk and the number moving between categories during each reassessment, owing to genuine improvement or deterioration in status (i.e. factoring out changes due to improved knowledge or revised taxonomy). Values range from 1 (if no species face imminent extinction) to 0 (if all are extinct). By calculating Red List Indices for species triggering identification of protected IBAs and AZEs and comparing to indices for those triggering unprotected sites, it is possible to quantify differences associated with protection.

We examined recent Red List Index trends for 4445 bird species of global conservation significance for which IBAs have been identified and for 845 birds, mammals and amphibians for which AZEs have been identified. We found that the increase in extinction risk over the last two decades was half as large for bird species with >50% of the IBAs at which they occur completely covered by protected areas, compared with species with ≤50% of IBAs completely covered ($P < 0.0001$), and a third lower for birds, mammals and amphibians restricted to protected AZEs, compared with those restricted to unprotected or partially protected AZEs ($P < 0.0001$) (Figure 12.7). The observed trends for species differed significantly from those expected if protection of sites was assigned at random. Increases in extinction risk for species occurring in protected sites were significantly smaller than the distribution of values derived after randomly assigning species (in the observed proportions) as having >50% or ≤50% of IBAs protected, or completely protected versus incompletely/unprotected AZEs and repeating this 10,000 times ($P < 0.05$) (Figure 12.8).

It is unlikely that this result was simply because less-threatened sites may be more likely to be protected (Pressey et al., 2002; Joppa & Pfaff, 2009), as the result for IBAs held even when excluding non-threatened species with an annual % index

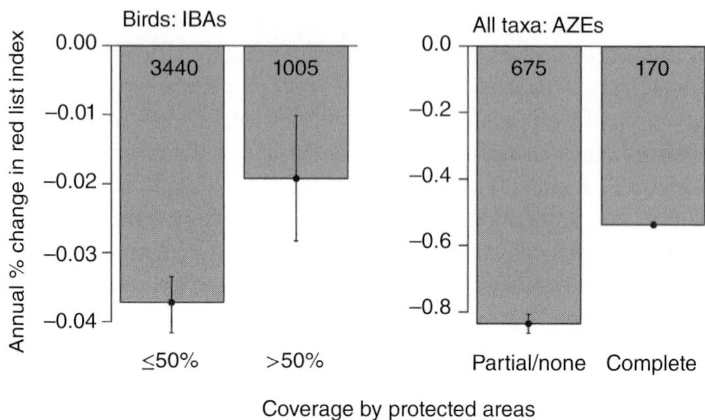

Figure 12.7 **Annual percentage decline in Red List Index for sets of bird species (during 1988–2008) with ≤50% or >50% of IBAs completely protected and for bird (1988–2008), mammal (1996–2008) and amphibian species (1980–2004) restricted to single sites (AZEs) that are partially/unprotected versus completely protected (averaged across taxa, weighting species equally). Numbers within each bar refer to the number of species. Error bars show 95% confidence intervals based on uncertainty around the estimated value that is introduced by data-deficient species. Source: Reproduced with permission from Butchart et al. (2012)**

decline = 0.186 for species with ≥50% of IBAs completely protected versus 0.251 for species with <50% sites protected (P = 0.044). In other words, the result cannot be explained by a hypothesis that protected sites disproportionately support non-threatened species, while unprotected sites disproportionately support threatened species. Furthermore, all AZEs are, by definition, under intense pressure, supporting the entire or overwhelming majority of the global population of at least one highly threatened species (Ricketts et al., 2005). Yet we still found an association between the degree of protection of these sites and reduction in Red List Index decline for species restricted to them. Finally, we found only a weak negative relationship between proportion of IBAs protected and local human population density (which is likely to be correlated with intensity of pressures) across all countries and levels of economic development ($F_{1,9114} = 4.74$, $P = 0.03$, slope ± SE = −0.028 ± 0.013), although this was stronger when the analysis was restricted to developing countries ($F_{1,4286} = 14.25$, $P = 0.0002$, slope ± SE = −0.083 ± 0.022).

The Red List Index is a useful approach for examining trends in the extinction risk of species, synthesising information on changes in species' population size, structure and trends and in their extent of distribution into a single index of aggregate survival probability (i.e. the inverse of extinction risk). As the system of Red List Categories and Criteria are designed to deal with uncertainty and paucity of information, they can be applied to all species globally within a taxonomic group, including poorly known tropical species (albeit with extremely poorly known

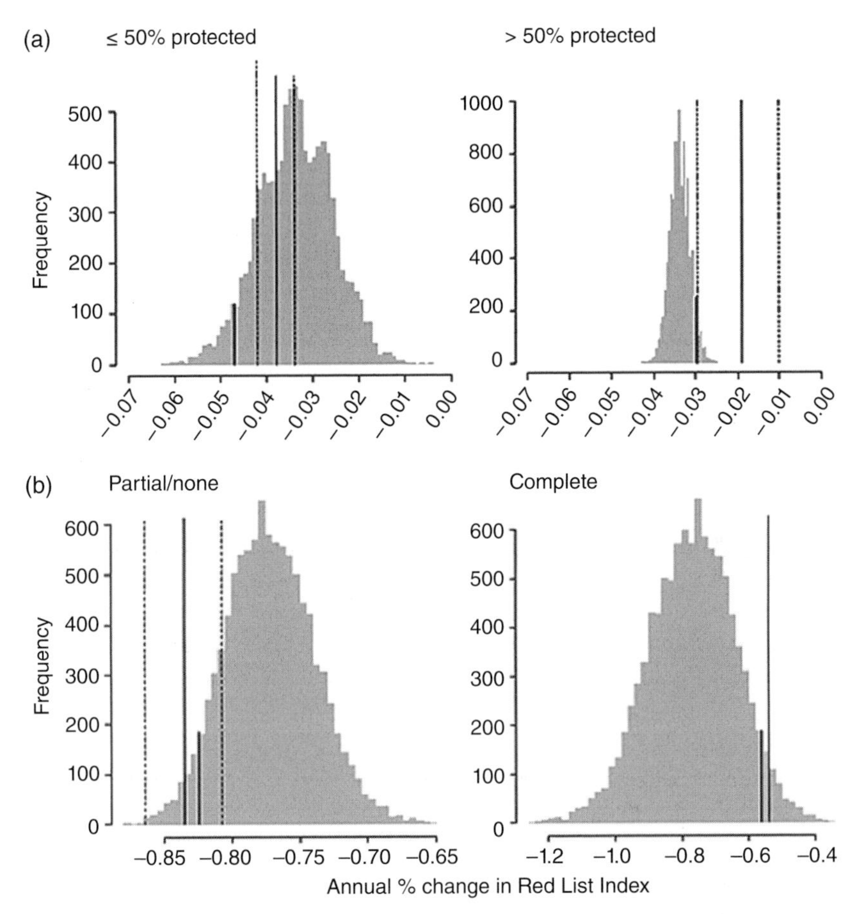

Figure 12.8 **Observed annual percentage declines in Red List Index (RLI) are significantly different from those expected by chance based on 10,000 randomisations for (a) bird species (during 1988–2008) with >50% of IBAs completely protected (N=1004, P<0.001) and (b) for bird (1988–2008), mammal (1996–2008) and amphibian species (1980–2004) restricted to single sites (AZEs) that are partially/unprotected (N=675, P=0.025) versus completely covered by protected areas (N=170, P=0.032). The RLI for bird species with ≤50% of IBAs completely protected was not significantly different from random (N=3440, P=0.31 (a)). The observed annual percentage change in RLI is shown as black lines (with 95% confidence intervals based on uncertainty introduced by data-deficient species shown by dashed lines) and annual percentage change in RLI from randomly allocating species 10,000 times shown by grey bars, with short black lines indicating the 5% confidence interval for a one-tailed test. Source: Reproduced with permission from Butchart et al. (2012)**

species classified as data deficient). This allows comparisons to be made in the broad trends in extinction risk for different subsets of species globally. However, owing to the breadth of each Red List Category, the Red List Index is only moderately sensitive. It is likely that many species in the taxonomic groups we considered experienced increases or decreases in extinction risk during the period but insufficiently to cross thresholds for higher or lower Red List Categories. Such trends are therefore not reflected in the index. Similarly, widespread abundant species that are classified as least concern may have declined in population size by up to 25% or may have increased substantially without such changes being reflected in revised Red List Categories and hence incorporated into the Red List Index.

Assessing the impact of protected areas on biodiversity trends would in theory be best achieved through comparing population trends of target species within pairs of protected and unprotected sites that are matched to control for potentially confounding variables (size, location, human influence, etc.). However, population trend data representatives of individual sites are scarce for most taxa, even in the best-studied groups like large mammals and birds and particularly in the tropics (Collen et al., 2008). Given this constraint, use of a metric at the level of species (rather than of populations in individual sites) and of aggregate extinction risk across species (rather than population trends per se) is sensible. It is reasonable to assume that adequate protection of the sole site harbouring the last remaining population of a species (in the case of AZEs) or of a suite of sites identified as the most important for the conservation of a species (IBAs) would affect the population trends, habitat extent and condition, sufficiently to influence its IUCN Red List Category of extinction risk. Hence, the Red List Index provides a useful tool for detecting moderately substantial differences in extinction risk trends for species occurring at sets of sites with different degrees of protection. It is consequently most useful in this way when applied at large spatial scales.

Conclusions and policy implications

Target 11 in the CBD Strategic Plan on Biodiversity calls for parties to conserve 17% of terrestrial (and inland water) areas by 2020, 'especially areas of particular importance for biodiversity' (CBD, 2010). IBAs and AZEs represent global networks of sites that are identified on the basis of current knowledge as the most important places for the persistence of global biodiversity (Ricketts et al., 2005; BirdLife International, 2010). Expansion of protected area networks and other site safeguard mechanisms to cover all partially protected or unprotected AZEs (459) and IBAs (8106) would add a further 4.6 million km^2, increasing terrestrial coverage from 12.9 to 17.5%. This would meet the 17% coverage target.

Within individual countries (the scale at which decisions about precisely which areas to protect and manage for biodiversity), a 'return on investment' approach could be used to determine the most efficient way to incorporate protection of important sites into protected area networks, given a fixed budget, particular land costs and specified biodiversity objectives (Murdoch et al., 2007, 2010; Watson et al., 2010).

An additional CBD Target 12 calls for the 'extinction of known threatened species to be prevented and their conservation status, particularly of those most in decline, to be improved and sustained by 2020' (CBD, 2010). Effective conservation of all AZEs is by definition essential to achieve this target as all such sites are under threat and the loss of any one of them in the short to medium term would almost certainly result in global extinction of at least one species (Ricketts et al., 2005). It is highly likely that IBAs are the most urgent priorities for conservation action to achieve the CBD target of improving the status of known threatened species (CBD, 2010), at least for birds. It would also make a substantial contribution for other taxa, given that IBAs comprise over 70% of the number and 80% of the area of important sites for mammals, amphibians and certain reptile, fish, plant and invertebrate clades (Butchart et al., 2012).

While there has been considerable progress in expanding the global protected area network in recent decades, this has not delivered adequate coverage of important sites for species conservation. In order to meet CBD Targets for 2020, as well as the Millennium Development Goals for 2015, for which protected area coverage of IBAs/AZEs and Red List Indices are used as indicators (United Nations, 2010), new and expanded protected areas need to target better the existing and systematically identified global networks of important sites. Furthermore, most protected areas need enhanced management in order to conserve biodiversity effectively in the long term, in the face of intense pressures. Application of the state–pressure–response indicators framework as exemplified by IBA monitoring can contribute to assessing the effectiveness of such management.

Acknowledgements

We are grateful to our co-authors on the papers upon which this chapter draw, as well as the thousands of individuals and organisations, including BirdLife International Partners, who contribute to Red List assessments of the world's birds, identification of IBAs and monitoring of their status (particularly for the Kenyan IBAs discussed here), plus to the AZE member organisations, contributors to the identification of AZEs and contributors to the WDPA. Partial support for this study was provided by an award from the John D. and Catherine T. MacArthur Foundation to NatureServe addressing 'Dashboard Assessments: Proof-of-Concept and Baselines'.

References

American Bird Conservancy, 2013. Alliance for Zero Extinction. [Online] Available at: http://www.zeroextinction.org/search.cfm [Accessed 7 August 2013].

Bennun, L. & Njoroge, P., 1999. *Important bird areas in Kenya*. Nairobi: Nature Kenya.

Bennun, L. et al., 2005. Monitoring important bird areas in Africa: Towards a sustainable and scalable system. *Biodiversity Conservation*, 14, p. 2575–2590.

BirdLife International, 2006. *Monitoring important bird areas: A global framework*. Cambridge: BirdLife International.

BirdLife International, 2010. State of the world's birds: Indicators for our changing world. [Online] Available at: http://www.birdlife.org/datazone/sowb [Accessed 1 October 2010].

Brooks, T. M. et al., 2006. Global biodiversity conservation priorities. *Science*, 313, p. 58–61.

Butchart, S. H. M. et al., 2004. Measuring global trends in the status of biodiversity: Red list indices for birds. *PLoS Biology*, 2(12), p. 2294–2304.

Butchart, S. H. M. et al., 2005. Using red list indices to measure progress towards the 21010 target and beyond. *Philosophical Transactions of the Royal Society of London Series B*, 360(1454), p. 255–268.

Butchart, S. H. M. et al., 2007. Improvements to the red list index. *PLoS ONE*, 2(1), e140.

Butchart, S. H. M. et al., 2012. Protecting important sites for biodiversity contributes to meeting global conservation targets. *PLoS ONE*, 7(3), e32529.

CBD, 2010. *Decisions adopted by the CoP to the CBD at its 10th meeting (UNEP/CBD/COP/DEC/X/2)*. Montreal: Secretariat of the CBD.

Chan, S., Crosby, M. J., Islam, M. Z. & Tordoff, A. W., 2004. *Important bird areas in Asia: Key sites for conservation*. Cambridge: BirdLife International.

Collen, B., Ram, M., Zamin, T. & McRae, L., 2008. The tropical biodiversity data gap: Addressing disparity in global monitoring. *Tropical Conservation Science*, 1(2), p. 75–88.

Davenish, C. et al., 2009. *Important bird areas in the Americas: Priority sites for biodiversity conservation*. Quito: BirdLife International.

Eken, G. et al., 2004. Key biodiversity areas as site conservation targets. *BioScience*, 54(12), p. 1110–1118.

Evans, M. I., 1994. *Important bird areas in the Middle East*. Cambridge: BirdLife International.

Evans, M. I. & Heath, M. F., 2000. *Important bird areas in Europe: Priority sites for conservation*. Cambridge: BirdLife International.

Fishpool, L. D. C. & Evans, M. I., 2001. *Important bird areas in Africa and associated islands: Priority sites for conservation*. Cambridge: BirdLife International.

Fishpool, L. D. C. et al., 1998. Important bird areas: Criteria for selecting sites of global conservation significance. *Proceedings of 22 International Ornithological Congress, Duban, South Africa*, 16–22 August, Issue 428.

Holland, R. A., Darwall, W. R. T. & Smith, K. G., 2012. Conservation priorities for freshwater biodiversity: The key biodiversity area approach refined and tested for continental Africa. *Biological Conservation*, 148(1), p. 167–179.

Joppa, L. N. & Pfaff, A., 2009. High and far: Biases in the location of protected areas. *PLoS ONE*, 4(12), e8273.

KNBS, 2010. *Kenya demographic and health survey 2008–2009*. Calverton: KNBS & ICF Macro.

Langhammer, P. F. et al., 2007. *Identification and gap analysis of key biodiversity areas: Targets for comprehensive protected area systems*. Gland: IUCN.

Luck, G. W., 2007. A review of the relationship between human population density and biodiversity. *Biology Reviews*, 82, p. 607–645.

Murdoch, W. et al., 2007. Maximising return on investment in conservation. *Biological Conservation*, 139(3), p. 375–388.

Murdoch, W., Ranganathan, J., Polasky, S. & Regetz, J., 2010. Using return on investment to maximise conservation effectiveness in Argentine grasslands. *Proceedings of the National Academy of Sciences of the United States of America*, 107(49), p. 20855–20862.

Musila, S. N. et al., 2006. *Kenya's important bird areas status and trends 2005*. Nairobi: Nature Kenya.

Mwangi, M. A. K. et al., 2010. Tracking trends in key sites for biodiversity: A case study using important bird areas in Kenya. *Bird Conservation International*, 20, p. 215–230.

NES, 1994. *The Kenya national environment action plan (NEAP)*. Nairobi: The National Environment Secretariat.

NES, 2000. *The Kenya national biodiversity strategy and action plan*. Nairobi: The National Environment Secretariat.

Ng'weno, F., Matiku, P. & Otieno, N., 2004. *Kenya's important bird areas: Status and trends 2004*. Nairobi: Nature Kenya.

PlantLife International, 2004. *Identifying and protecting the world's most important plant areas: A guide to implementing target 5 of the global strategy for plant conservation*. Salisbury: PlantLife International.

Pressey, R. L., Whish, G. L., Barrett, T. W. & Watts, M. E., 2002. Effectiveness of protected areas in north eastern New South Wales: Recent trends in six measures. *Biological Conservation*, 106(1), p. 57–69.

Ricketts, T. H. et al., 2005. Pinpointing and preventing imminent extinctions. *Proceedings of the National Academy of Sciences of the United States of America*, 102(51), p. 18497–18501.

Salafsky, N. et al., 2008. A standard lexicon for biodiversity conservation: Unified classifications of threats and actions. *Conservation Biology*, 22(4), p. 897–911.

Sparks, T. H. et al., 2011. Linked indicator sets for addressing biodiversity loss. *Oryx*, 45, p. 411–419.

United Nations, 2010. *The millennium development goals report 2010*. New York: UN DESA.

Van Swaay, C. A. M. & Warren, M. S., 2003. *Prime butterfly areas in Europe: Priority sites for conservation*. Wageningen: National Reference Centre of Agriculture, Nature Management and Fisheries.

Watson, J. E. M. et al., 2010. The capacity of Australia's protected area system to represent threatened species. *Conservation Biology*, 25, p. 324–332.

13

Camera Traps for Conservation: Monitoring Protected Area Investments

T. G. O'Brien

Wildlife Conservation Society, Bronx, NY, USA

Introduction

The Conference of Parties to the Convention on Biological Diversity (CBD) has recognized the need and importance of monitoring to assess global progress on meeting the 2020 Aichi Targets. They have been proactive in developing indicators to measure progress in meeting targets but recognize that sources and nature of data may be problematic (Balmford et al., 2005; Dobson, 2005; CBD, 2010). Most indicators use available data that are several years old due to the lag between data collection and publication. These data are overwhelmingly concentrated in the temperate zone rather than the tropics and often lack adequate quality control over accuracy and precision. Progress toward achieving 2020 targets will be difficult to ascertain if the indicator relies on old data, is mostly from the biologically less diverse temperate region, and is of questionable accuracy. An indicator is only as good as its underlying data.

The Strategic Goal C of the Aichi 2020 objective sets a lofty target, "To improve the status of biodiversity by safeguarding ecosystems, species and genetic diversity" (CBD, 2010) by the end of the decade. The targets for Strategic Goal C include Target 11, "By 2020, at least 17% of terrestrial and inland water, and 10% of marine areas, especially areas of importance for biodiversity and ecosystem services, are conserved through effectively and equitably managed, ecologically representative and well-connected systems of protected area and other effective area-based conservation measures, and integrated in the wider landscapes and seascapes" (CBD, 2010), and Target 12, "By 2020, the extinction of known threatened species

Protected Areas: Are They Safeguarding Biodiversity?, First Edition.
Edited by Lucas N. Joppa, Jonathan E. M. Baillie and John G. Robinson.

has been prevented and their conservation status, particularly of those most in decline, has been improved and sustained" (CBD, 2010).

Protected areas have always been an integral part of CBD's efforts to conserve biodiversity, and these targets solidly reaffirm the commitment to place protected areas and protected area management at the heart of the latest strategy to conserve biodiversity. The two headline indicators for Target 11 are "coverage of protected areas and protected area management," while the two headline indicators for Target 12 include "change in status of threatened species" and "trends in abundance and distribution of selected species" (Table 13.1). The proposed primary indicators to support the headline indicators all require that new data be collected to address the status of the indicator and that a strategy for future data collection be put in place (CBD, 2011).

Targets 11 and 12 make clear that the key to making protected areas more effective is improved management and monitoring of management interventions. Adaptive management requires that we make an intervention, monitor the outcome, review the results of monitoring, modify our interventions, and apply again. In order to be effective in monitoring, we need monitoring metrics that accurately reflect the situation in the field. We need to move beyond expert opinion.

To demonstrate that management interventions are effective, we need to monitor (WCS, 2002; Strindberg & O'Brien, 2012):

- The conservation interventions themselves (performance monitoring)
- The changes in the influence of threats (outcome monitoring)
- The effect on conservation targets (impact monitoring)

Performance monitoring is important to be able to assess if management actions are actually being taken. Outcome monitoring is important to assess whether the interventions are changing the status of the threats. And impact monitoring is the ultimate indicator, as it identifies whether there is an improvement in the state of the conservation target itself. Such improvements give us the greatest level of confidence that we might be achieving our goal, but monitoring conservation targets is often the most difficult, costly, and time-delayed form of monitoring.

Monitoring interventions and threats (both of which have relatively low cost and rapid results) may act as a proxy for monitoring conservation targets, but there are trade-offs. These indicators can fluctuate through time and provide a lower level of confidence that the data are providing meaningful information about actual conservation success. The proposed indicators for Target 11 (Table 13.1) are primarily monitoring the establishment of protected areas. Only one primary indicator, "protected area management effectiveness," is monitoring changes in the state of threats, and none are measuring the efficacy of interventions to protect biodiversity. Given the huge uncertainties associated with management, due to a lack of understanding of ecological processes, environmental variation, exploitation, and our imperfect ability to observe the state of nature accurately,

Table 13.1 **CBD Aichi 2020 targets relating to protected areas and biodiversity**

Target	Headline indicator	Primary indicators	Level of monitoring
11: By 2020, at least 17% of terrestrial and inland water, and 10% of marine areas, especially areas of importance for biodiversity and ecosystem services, are conserved through effectively and equitably managed, ecologically representative and well-connected systems of protected area and other effective area-based conservation measures, and integrated in the wider landscapes and seascapes	Coverage of protected areas	Coverage of terrestrial protected areas	Intervention
		Proportion of ecoregions protected	Intervention
		Coverage of marine protected areas	Intervention
		Coverage of inland water protected areas	Intervention
		Proportion of key biodiversity areas protected	Intervention
		Number of protected areas with connectivity corridors and buffer zones	Intervention
	Protected area management	Number of countries with complete ecological gap analysis	Intervention
		Protected area management effectiveness*	Intervention/threat
12: By 2020, the extinction of known threatened species has been prevented and their conservation status, particularly of those most in decline, has been improved and sustained	Change in status of threatened species	Red List Index	Outcome
	Trends in abundance and distribution of selected species	Living Planet Index	Outcome
		Global Wild Bird Index*	Outcome
		Arctic Species Trend Index*	Outcome
		Waterbird Population Status Index*	Outcome

Source: Data from CBD (2011).
*Primary Indicator under development.
Each target is accompanied by a proposed headline indicator and supporting primary indicators.
The level of monitoring (Intervention, Threat, Outcome) is in order of increasing complexity.

these indicators will not reliably indicate whether Target 11 has been achieved. The area covered by protected areas does not, by itself, provide a good indication of conservation. The long-standing critique of "paper parks" (Dudley & Stolton, 1999) is a poignant illustration of the gap between legally establishing protected area and the efficacy of mandated protection.

Species level indicators

To answer the question of whether protected areas help to achieve Target 12 by maintaining intact wildlife communities, we need to know if the management interventions are:

- Preventing local extinction
- Promoting recovery and colonization of threatened species
- Retaining stable or expanding populations of birds and mammals

Target 12 indicators, in contrast to Target 11 indicators, focus on conservation targets of management interventions.

The Red List Index (RLI) (Butchart et al., 2004, 2005) measures the trends in extinction risk of species, based on the number of species moving between risk categories. RLI can be aggregated across taxa to develop a biodiversity extinction risk index. Assessments occur every 5 years and changes occur only if supporting data are available. Data required for a change in risk classification include species' abundance, geographic distribution, and rate of decline. The RLI therefore is reliant on timely dissemination of data into the public domain, and the data need to be of sufficient quality to satisfy IUCN Red List Criteria.

The Living Planet Index (LPI) (Loh et al., 2005; Collen et al., 2009) measures the aggregated trend in 9014 populations of 2688 species of mammal (464), bird (1320), reptile, amphibian, and fish species in different biomes in the temperate and tropical zones (WWF, 2012). Changes in populations over time are scaled and aggregated for the temperate and tropical zones, and the geometric mean of the temperate and tropical region is then calculated. Data underlying the LPI include population size, density, distribution, and other proxies of abundance. Data are only included if a measure or index of population size is available for at least 2 years and information is available on how data were collected, what the units of measurement were, and the geographic location of the population. Data must be collected using the same method on the same population throughout the time series, and data sources must be referenced and traceable. The index is approximately 50% bird species.

The Global Wild Bird Index (GWBI) is still in development and will measure population trends of a representative suite of wild birds as an indicator of environmental health. The methods for producing bird indices are well developed, and wild bird indices have already been produced for Europe and North America (Gregory et al., 2003, 2005). The GWBI measures biodiversity change in a similar fashion to the LPI; the main difference is that the GWBI only incorporates trend data from formally designed breeding bird surveys to deliver scientifically robust and representative indicators. The requirement for robust data, however, means that data coverage is currently patchy and the GWBI is not presently applicable at a global scale.

The Waterbird Population Status Index (WPSI) measures the trend in water bird populations globally. Unlike LPI and WBI, the WPSI is based on the direction of the trend rather than the magnitude of change (Wetlands International, 2010). It has the advantage of good coverage in the tropics and has standardized data collection protocols.

The Arctic Species Trend Index (ASTI) (Bohm et al., 2012) measures the trend in 323 arctic species (37% of all species) including all the mammals, more than half of the birds, and 27% of fish species. Part of the data overlap the LPI and bird populations account for more than half of the populations. Like the WPSI, this index reports patterns in trends of populations.

As is clear from this brief review, the Target 12 primary indicators are predominately based on bird populations. In part, this is because birds are relatively simpler to sample in many habitats than other taxa and because international bird conservation efforts are better organized. Birdwatching has a wide following globally and many countries maintain organized bird monitoring programs, either through NGO activities or the government. Other taxonomic groups are not as easily censured as birds, due to cryptic nature and often nocturnal habits. Amphibians and reptiles, in particular, rarely are the subject of long-term monitoring programs. Mammals too are underrepresented in monitoring programs.

In the remaining sections, I review the kinds of monitoring activities typically conducted in protected areas and how well the resulting data might be used for reporting against Aichi Targets 11 and 12. I then recommend an approach, monitoring using camera traps, which allows for cost-effective, consistent, and scalable monitoring of multiple taxa across space and time.

Scaling biodiversity surveys to monitor Aichi progress: Field surveys and camera trapping

Effective monitoring of birds and mammals in protected areas is a daunting challenge. Many protected areas are in remote areas where conducting monitoring can require complex and often expensive logistics. Bird communities can be incredibly complex, with many species that are uncommon or rare, often cryptic, occasionally nocturnal, and inhabiting all strata of forests and other habitats. Mammals too inhabit all strata of a habitat, usually avoid humans, and are often rare, cryptic, and nocturnal. While we have made great strides in remote-sensing monitoring of ecosystems, we have made less progress in monitoring wildlife diversity accurately within those ecosystems. Most wildlife monitoring still relies on natural history observation skills, sometimes accompanied by clever sampling schemes. We search for sign, listen for calls, walk miles and miles to encounter wildlife and use ingenious methods to estimate abundance and density from encounters with dung or using DNA from hair. This requires people spending long hours working in the field counting species, individuals, and their sign. Indeed, the CBD and Aichi 2020 targets require that we count species and individuals if we are to establish that wildlife communities are recovering or are stable.

Most protected area monitoring for wildlife takes the form of:

- Transects (Buckland et al., 2010), "recces" (Walsh & White, 1999), or law enforcement patrols in which wildlife observations and sign observations also are recorded (Stokes, 2010)
- Interviews with park staff or local villagers about protected area effectiveness (Poulsen & Luanglath, 2005)
- Unstructured reports by people visiting protected areas for a variety of reasons (tourism, nontimber forest product extraction, hunters)
- Expert opinion
- Analysis of budgets to determine support for monitoring

The use of interviews, expert opinion, and budget allocations may be sufficient for monitoring interventions and threats, but only field-based monitoring activities are adequate for measuring impacts on conservation targets.

The most common forms of field-based monitoring include informal reporting of wildlife sightings by tourists, community-based monitoring, security patrols, animal/animal sign transects, and line transect surveys (aerial and ground). These methods are almost exclusively employed for daytime sampling; nocturnal animals are usually monitored indirectly, using sign surveys. Most of the methods involve walking established trails, following the path of least resistance off trails, driving roads, and observing at points like water holes and salt licks where wildlife congregate. With the exception of properly designed line transect surveys (Buckland et al., 2010), these methods result in some form of catch-per-unit-effort (CPUE) metric, a form of relative abundance index (O'Brien, 2011). The CPUE index is then assumed to reflect real changes in species abundance. The index does not incorporate relative detectability of animals being counted nor does it attempt to estimate or incorporate the undetected fraction of wildlife. Use of CPUE indices requires assumptions about wildlife behavior, human behavior, and the nature of biases to be valid. First, we assume that all species are distributed uniformly over the landscape and that group sizes of each species are also distributed randomly over the landscape. Second, we assume that the sampling methods are strictly standardized such that variation in the application of sampling methods does not affect the index. Third, we assume that any bias inherent in the index (i.e., detection or visibility bias) is constant over space and time.

To satisfy the conditions of the first assumption, surveys should sample the area of interest in a randomized or systematic manner in order to have an equal probability of encountering animals (and groups of different sizes) throughout the study area. To satisfy conditions of the second assumption, the execution of surveys must be rigidly standardized and monitored, environmental conditions standardized, and skill levels of observers known and controlled. Conditions necessary to satisfy the third assumption are so wide ranging that they are difficult to describe in detail, except that any conditions not covered by the first and second assumption that might affect the detection of sign or animals must be constant over time. Detection may change over space and time for a wide variety of reasons and many not apparent to the investigator. The failure of the third assumption (constant bias) can result in mixing changes in abundance and changes in detection and result in incorrect interpretation of the CPUE index. If CPUEs are effective indices of abundance, we would expect that over time the relationship between the CPUE index and estimated density would be monotonic, positive, and highly correlated. A CPUE that is only weakly correlated with abundance is of little use as a monitoring metric. The only way to know if an encounter rate index accurately reflects population abundance is to calibrate the index with independently derived, unbiased estimates of abundance or density (O'Brien, 2011; Kinnaird & O'Brien, 2012) or to model the underlying process that generates the encounter rate (Jennelle et al., 2002; Rowcliffe et al., 2008). Unfortunately, many protected area monitoring programs use abundance indices that have not been evaluated for accuracy (e.g., Danielsen et al., 2000; Poulsen & Luanglath, 2005; Cassidy, 2007).

When signs are incorporated into encounter rates, we have additional problems of misidentification and double counting. Interpreting sign correctly requires ancillary data on deposition rates, decay rates (feces), and excellent natural history skills. Because sign surveys are most often conducted along trails, assumptions about how animals are distributed in space and how they move through their habitats become critically important. Sign surveys may be adequate to establish the presence of a species but are not useful for monitoring abundance and species richness except under exceptionally controlled conditions (Gopalaswamy et al., 2012).

It has become increasingly expensive to conduct monitoring activities in remote protected areas globally. The gross national income per capita (GNI), a measure of the average income in a country, has grown by more than 280% globally between 1998 and 2012 (World Bank, 2014). Methods of remote monitoring that reduce labor costs are becoming cost-effective alternatives to sending field teams into remote areas for long periods to collect observation data.

Acoustic sensors to monitor diversity of vocalizations (bats, amphibians, birds, nocturnal mammals) (Wrege et al., 2012; Jones et al., 2013) are developing rapidly. Vocalizations are a special form of sign survey. When we know the calling rate and effective sampling area, call counts can be converted to density. Currently, acoustic sensors are most useful for species richness assessments. It is still difficult to know the call rates for an entire mammal or bird community and to calculate an effective sampling area for an acoustic sensor. Acoustic methods have not yet been used for diversity monitoring of terrestrial mammals.

An alternative to diurnal field surveys for monitoring mammals and terrestrial birds is the application of camera traps (Kinnaird & O'Brien, 2012; O'Brien & Kinnaird, 2013). Use of camera traps as a monitoring tool has increased exponentially since they were first introduced in the 1990s (Figure 13.1) (Rowcliffe & Carbone, 2008). Between 1998 and 2011, the average camera trap cost dropped from approximately $400 per unit to $210 per unit. Meanwhile, the performance of camera traps has increased from 36 photo film images to 20,000+ digital images, and battery life in the field has increased from 1 month to potentially 1 year (Figure 13.1). Many models are weather resistant with high heat and cold tolerance, allowing use in wide range of environmental conditions including humid tropical rainforests and Antarctica.

Camera traps have a number of distinct advantages as data-collecting units. First, camera traps provide photographic evidence of the species and number of individuals detected and the evidence can be verified by multiple experts. Misidentification of species is problematic in all forms of multispecies sampling, but without a photograph or vocal recording, problematic identifications cannot be checked and verified once the observer has left the field. Second, camera traps are active 24 hours a day and pick up a great variety of diurnal, crepuscular, and nocturnal species, increasing the breadth of biodiversity inventories. In a recent camera trap survey in Laikipia, Kenya, Kinnaird and O'Brien (unpublished) recorded 62 mammal species of which 40% were nocturnal. During camera trap surveys in six Southeast Asian countries, 44% of the detected species were considered rare, occurring less than once per thousand trap days (O'Brien, unpublished). Third, interobserver reliability is high; camera traps are built to specifications and are more likely to perform similarly compared to multiple human observers. Finally, properly designed camera trap monitoring programs can generate monitoring metrics for entire communities that are unbiased, removing the major source of uncertainty about trends in biodiversity.

Camera trap surveys can provide valuable data on three important state variables: abundance, distribution and species richness, that are important for interpreting trends in diversity. Camera traps are typically used to estimate abundance and density of a target species when its members are recognizable as individuals (O'Brien, 2011), to estimate density (Rowcliffe et al., 2008) or point abundance (Royle & Nichols, 2003) of a species when individuals are not recognizable as individuals, to estimate relative abundance of a species using accumulation rates of independent photographic events (O'Brien et al., 2003; Rovero & Marshall, 2009), to estimate species richness of a community (Kery, 2011; O'Brien et al., 2011), and to estimate the proportion of area occupied by a species or group of species (MacKenzie et al., 2006; O'Connell & Bailey, 2011). Camera traps can also be used to monitor species or community dynamics over time. Sampling designs for

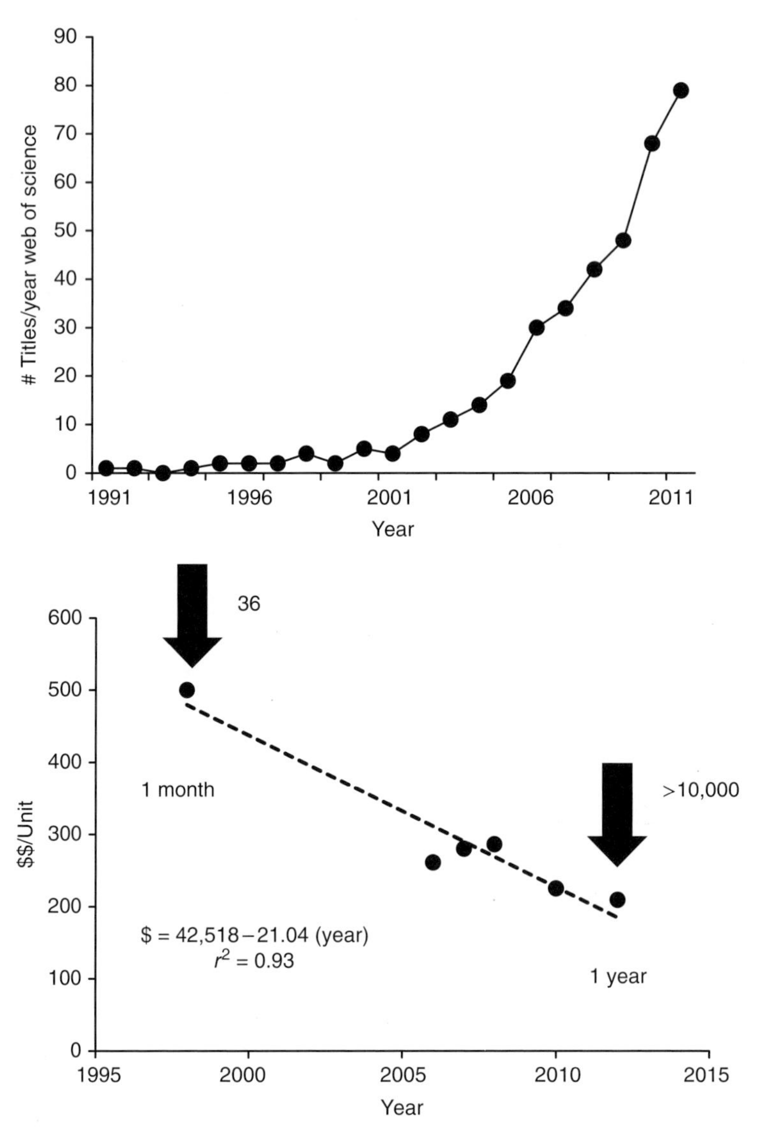

Figure 13.1 **Growth in research publications citing use of camera traps to collect data (Source: Web of Science), falling cost of camera traps (closed circles, regression line), increasing capacity (images on film versus digital cameras: above line), and increased battery life (below line) (Reuters, 2015)**

deployment of cameras may vary for different objectives (maximizing detection probabilities versus optimizing coverage) and some metrics require additional data to evaluate underlying assumptions (relative abundance indices based on photograph accumulation rates).

An example of large-scale application of camera trapping as a monitoring method in tropical rainforests is provided by The Tropical Ecology Assessment and Monitoring (TEAM) program (TEAM Network, 2011), a consortium composed of Conservation International, Smithsonian Institution, and Wildlife Conservation Society. TEAM's mission is to generate real-time data for monitoring long-term trends in tropical biodiversity through a global tropical network, providing an early warning system on the status of biodiversity to guide conservation action. The consortium has been functioning for 6 years and includes 16 rainforest sites located in protected areas of 15 countries. TEAM works closely with more than 80 institutional partners, including national governments and local nongovernmental organizations (NGOs). An important component to TEAM monitoring has been the development and implementation of standardized camera trapping protocols at all sites (TEAM Network, 2011). These sites currently monitor 341 species of terrestrial birds and mammals; more than 50% of the total species count for the LPI Tropical Terrestrial Index. A broader application of a TEAM-style camera trap monitoring program could potentially provide information on up to 2000 species of birds and mammals annually (O'Brien & Kinnaird, 2013).

The advantages of using camera trap in a monitoring network are many. First, the skill set necessary to execute a camera trap sampling design is reasonably simple, requiring knowledge of the operation of the camera traps, use of Global Positioning System (GPS), ability to navigate across a landscape, and the implementation of standard logistics relevant to the site. No special skills in animal identification or sign interpretation are necessary and this facilitates standardization. Second, because data collection is automated and standardized across sites, monitoring metrics (i.e., species richness) can be calculated without having to assess whether the underlying data and sampling efforts are comparable. Third, the data can be collected on a standardized schedule (annual or biannual) and incorporated into annual work plans. Finally, because the data are collected automatically, a two-month deployment of camera traps requires less than 2 months of fieldwork by the monitoring team, a savings on the ever-increasing cost of labor. Data interpretation is straightforward, involving reading exchangeable image file format (Exif) metadata on the images and identifying species. Although misidentification is still a problem with camera trap data, it is possible to consult experts for confirmation, which is not possible when the data consist of written reports of observations without accompanying documentation.

Cost considerations for camera trap monitoring in protected areas

It is difficult to estimate the comparative costs of camera trapping and other forms of monitoring without actually conducting both kinds of sampling at a site and comparing the costs. O'Brien & Kinnaird (2013) used this approach to show that camera trap survey costs in African savannah and Asian rainforest were approximately 30 and 15% less than similar efforts using line transect surveys. The cost difference was attributed primarily to differences in salaries and logistics for transect surveys because transect sampling required that the monitoring team spend more days in the field.

In general, camera trap surveys and other kinds of transect-/patrol-based surveys have similar logistical challenges. Teams have to be transported and maintained in the field. Both logistics and salaries costs are a function of time spent in the field. Start-up costs are higher for camera trap surveys than most other kinds of surveys. Camera trap surveys require the camera units, memory cards, batteries, locks, and chargers. Line transect- and patrol-based surveys require binoculars, compasses, and range finders. If start-up costs are amortized over the useful life of the equipment, then transect survey start-up costs become trivial and the cost of camera traps becomes reasonable. So the key to cost-effectiveness of camera trap surveys compared to transect-based surveys is the actual cost of implementation versus the start-up cost.

Ahumada & O'Brien (2013) recently compared the cost of traditional observer-based monitoring and camera trap monitoring for Latin American protected areas. They used data from Vreugdenhil (2004) for the cost of protected area management in Latin America (adjusted to $2012) to estimate current effective protected area budgets and staff salaries (Vreugdenhil, 2004). Start-up equipment costs for camera trap surveys included US$180 (low-end) and US$440 (high-end) camera traps with SD cards (US$10), rechargeable batteries (US$30), and locks (US$20) for 50 cameras. Start-up costs for trail-based surveys included binoculars (US$250), laser rangefinders (US$350), and digital compasses (US$50).

They used camera trap performance data from 6 years of TEAM data to estimate the failure rate of camera traps at 1.2% per year (range 0.3–8.0%). They estimated that a camera trap has an average life of 10-month long deployments at the low end and perhaps 50-month long deployments at the high end (though this is conservative). A stock of 50 low-end camera traps deployed at a rate of 30 cameras per sample block might be expected to last for 18 deployments before replacement. A stock of 50 high-end camera traps would last for 74 deployments before replacement. They included replacement costs of cameras and amortized the cost over 10 years of deployment.

Ahumada & O'Brien (2013) estimated implementation costs as a function of data collection and data management. Field implementation costs assumed that survey teams for each protected area were composed of two rangers (or trained technicians) and three porters. Teams deployed 90 camera trap points or conducted 90 line transect surveys (2 surveys per day plus time to move). Other costs included standard per diems, transport, field rations, camping equipment, and minor equipment. Data management costs were estimated two ways. First, they assumed that each country acquires the hardware necessary to store and manage its data and the cost is amortized over the useful life of the hardware (5 years). Second, countries use global cloud solutions (e.g., Amazon, Google, etc.) for storage of images and data management. These options also included software maintenance and network costs and 1.5 months of an IT specialist salary in each country. They estimated the cost of a cloud-based data management strategy at US$4000 and the in-country server-based strategy at US$6250 annual costs. Camera trapping surveys were on average 10–20% less expensive than foot-based monitoring of similar intensity, and most of the cost difference was attributed to salaries and logistics of maintaining staff in the field for extended periods of time.

The increase in quantity and quality of data derived from camera trap monitoring is also substantial. O'Brien & Kinnaird (2013) found that camera trap surveys resulted in data on twice as many species as line transect surveys, due to detection of nocturnal and cryptic species that are easily missed on diurnal surveys. The combined cost savings and increase in information content argue powerfully for use of camera traps as a tool for monitoring biodiversity in protected areas.

As countries consider how to finance their commitments under the CBD and Aichi 2020 targets, with limited protected area budgets, a reasonable balance must be found between allocation to management interventions and to monitoring outcomes. A carefully designed monitoring program should result in accurate and precise metrics that are repeatable over time and should be affordable. Camera trapping is one of the few options for monitoring terrestrial birds and mammals that can yield relatively unbiased metrics for many species that are comparable over time and space and at a reasonable cost.

Acknowledgments

I would like to thank J. Ahumada, S. Andelman, J. Baillie, J. Ginsberg, K.U. Karanth, M. Kinnaird, J. Nichols, J. Robinson, and S. Strindberg for many years of discussion and insight into biodiversity conservation and monitoring. I would also like to thank J. Baillie, L. Joppa, and J. Robinson for inviting me to participate in this symposium and book.

References

Ahumada, J. A. & O'Brien, T. G., 2013. *Implementation of camera trap networks at the national level: A cost benefit analysis for the TEAM network experiences*, Gland: Unpublished to the CBD SBTTA.

Balmford, A. et al., 2005. The 2010 challenge: Data availability, information needs and extraterestrial insights. *Philosophical Transactions of the Royal Society of London Series B*, Volume 360, p. 221–228.

Bohm, M. et al., 2012. *Tracking trends in Arctic vertebrate populations through space and time*, Iceland: CAFF.

Buckland, S. T., Plumptre, A. J., Thomas, L., & Rexstad, E. A., 2010. Design and analysis of line transect surveys for primates. *International Journal of Primatology*, Volume 31, p. 833–847.

Butchart, S. H. M., Ackakaya, H. R., Kennedy, E., & Hilton-Taylor, C., 2005. Biodiversity indicators based on trends in conservation strengths of the IUCN red list index. *Conservation Biology*, Volume 20, p. 579–581.

Butchart, S. H. M. et al., 2004. Measuring global trends in the status of biodiversity: Red list indices for birds. *PLoS Biology*, Volume 2, p. 2294–2304.

Cassidy, L., 2007. *Management oriented monitoring system in southern African region: Where are we now and the way forward*, Kasane: WWF & DWNP.

CBD, 2010. *Decisions adopted by the CoP to the CBD at its 10th meeting (UNEP/CBD/COP/DEC/X/2)*, Montreal: Secretariat of the CBD.

CBD, 2011. *Ad hoc technical expert group: Indicators for the strategic plan for biodiversity 2011–2020 (UNEP/CBD/AHTEG-SP-Ind/1/2)*, Montreal: Secretariat of the CBD.

Collen, B. et al., 2009. Monitoring change in vertebrate abundance: The living planet index. *Conservation Biology*, Volume 23, p. 317–327.

Danielsen, F. et al., 2000. A simple system for monitoring biodiversity in protected areas of a developed country. *Biodiversity and Conservation*, Volume 9, p. 1671–1705.

Dobson, A., 2005. Monitoring global rates of biodiversity change: Challenges that arise in meeting the CBD 2010 goals. *Philosophical Transactions of the Royal Society of London Series B*, Volume 360, p. 229–241.

Dudley, N. & Stolton, S., 1999. *Conversion of paper parks to effective management: Developing a target*, Gland: IUCN & WWF.

Gopalaswamy, A. M., Karanth, K. U., Kumar, N. S., & Macdonald, D. W., 2012. Estimating tropical forest ungulate densities from sign surveys using abundance models of occupancy. *Animal Conservation*, Volume 15(6), p. 669–679.

Gregory, R. D. et al., 2003. Using birds as indicators of biodiversity. *Ornis Hungarica*, Volume 12(13), p. 11–24.

Gregory, R. D. et al., 2005. Developing indicators for European birds. *Philosophical Transactions of the Royal Society of London Series B*, Volume 360, p. 269–288.

Jennelle, C. S., Runge, M. C., & MacKenzie, D. I., 2002. The use of photographic rates to estimate density of tigers and other cryptic mammals: A comment on misleading conclusions. *Animal Conservation*, Volume 5, p. 119–120.

Jones, K. E. et al., 2013. Indicator bats program: A system for the global acoustic monitoring of bats. In: B. Collen, N. Pettorelli, J. E. M. Baillie, & S. M. Durant, eds. *Biodiversity monitoring and conservation: Bridging the gaps between global commitment and local action*. Chichester: John Wiley & Sons, Ltd., p. 213–248.

Kery, M., 2011. Species richness and community dynamics: A conceptual framework. In: A. F. O'Connell, J. D. Nichols, & K. U. Karanth, eds. *Camera traps in animal ecology: Methods and analysis*. Tokyo: Springer Verlag, p. 207–232.

Kinnaird, M. F. & O'Brien, T. G., 2012. Effects of private land use, livestock management and human tolerance on diversity, distribution and abundance of large African mammals. *Conservation Biology*, Volume 26(6), p. 1026–1039.

Loh, J. et al., 2005. The living planet index: Using species population time series to track trends in biodiversity. *Philosophical Transactions of the Royal Society of London Series B*, Volume 360(1454), p. 289–295.

MacKenzie, D. I. et al., 2006. *Occupancy estimation and modeling: Inferring patterns and dynamics of species occurrence*, Amsterdam: Elsevier.

O'Brien, T. G., 2011. Abundance, density and relative abundance: A conceptual framework. In: A. F. O'Connell, J. D. Nichols, & K. U. Karanth, eds. *Camera traps in animal ecology: Methods and analysis*. Tokyo: Springer Verlag, p. 71–96.

O'Brien, T. G. & Kinnaird, M. F., 2013. The wildlife picture index: A biodiversity indicator for top trophic levels. In: B. Collen, N. Pettorelli, J. E. M. Baillie, & S. M. Durant, eds. *Biodiversity monitoring and conservation: Bridging the gaps between global commitment and local action.* London: Wiley-Blackwell, p. 45–71.

O'Brien, T. G., Kinnaird, M. F., & Wibisono, H. T., 2003. Crouching tigers, hidden prey: Sumatran tiger and prey populations in a tropical forest landscape. *Animal Conservation*, Volume 6, p. 131–139.

O'Brien, T. G., Kinnaird, M. F., & Wibisono, H. T., 2011. Estimation of species richness of large vertebrates using camera traps: An example from an Indonesian rainforest. In: A. F. O'Connell, J. D. Nichols, & K. U. Karanth, eds. *Camera traps in animal ecology: Methods and analysis*. Tokyo: Springer Verlag, p. 233–252.

O'Connell, A. F. & Bailey, L. L., 2011. Inference for occupancy and occupancy dynamics. In: A. F. O'Connell, J. D. Nichols, & K. U. Karanth, eds. *Camera traps in animal ecology: Methods and analysis*. Tokyo: Springer Verlag, p. 191–206.

Poulsen, M. K. & Luanglath, K., 2005. Projects come, projects go: Lessons from participatory monitoring in southern Laos. *Biodiversity and Conservation*, Volume 14, p. 2591–2610.

Reuters, T., 2015. *Web of Science*. [Online] Available at http://apps.webofknowledge.com/ [Accessed August 7, 2013].

Rovero, F. & Marshall, A. R., 2009. Camera trapping photographic rates an index of density in forest ungulates. *Journal of Applied Ecology*, Volume 46, p. 1011–1017.

Rowcliffe, J. M. & Carbone, C., 2008. Surveys using camera traps: Are we looking to a brighter future. *Animal Conservation*, Volume 11, p. 185–186.

Rowcliffe, J. M., Field, J., Turvey, S. T., & Carbone, C., 2008. Estimating animal density using camera traps without the need for individual recognition. *Journal of Applied Ecology*, Volume 45, p. 1228–1236.

Royle, J. A. & Nichols, J. D., 2003. Estimating abundance from repeated presence-absence data or point counts. *Ecology*, Volume 84, p. 777–790.

Stokes, E. J., 2010. Improving effectiveness of protection efforts in tiger source sites: Developing a framework for law enforcement monitoring using MIST. *Integrative Zoology*, Volume 5, p. 363–377.

Strindberg, S. & O'Brien, T., 2012. A decision tree for monitoring wildlife to access the effectiveness of conservation interventions. Working paper 41, Bronx: WCS.

TEAM Network, 2011. Tropical Ecology Assessment & Monitoring Network. [Online] Available at: http://www.teamnetwork.org [Accessed August 7, 2013].

Vreugdenhil, D., 2004. Worldwide financing needs of protected area systems of developing and transition countries. *Conservacao e Natureza*, Volume 2(1).

Walsh, P. D. & White, L. J. T., 1999. What will it take to monitor forest elephant populations. *Conservation Biology*, Volume 13, p. 1194–1202.

WCS, 2002. *Monitoring conservation project effectiveness: Living landscapes bulletin 6*, Bronx: WCS.

Wetlands International, 2010. *State of the world's waterbirds 2010*, Ede: Wetlands International.

World Bank, 2014. GNI per Capita: Atlas Method. [Online] Available at: http://data.worldbank.org/indicator/NY.GNP.PCAP.CD/countries [Accessed August 7, 2013].

Wrege, P. H., Rowland, E. D., Bout, N., & Doukaga, M., 2012. Opening a larger window onto forest elephant ecology. *African Journal of Ecology*, Volume 50, p. 176–183.

WWF, 2012. *Living planet report 2012*, Gland: WWF.

Monitoring Protected Areas from Space

N. Pettorelli[1], *M. Wegmann*[2,3], *L. Gurney*[4] *and G. Dubois*[4]

[1]Institute of Zoology, Zoological Society of London, London, UK
[2]Remote Sensing and Biodiversity Research, University of Würzburg, Würzburg, Germany
[3]German Aerospace Centre, Cologne, Germany
[4]European Commission – Joint Research Centre, Brussels, Belgium

Introduction

Since the late 19th century, much of the conservation efforts carried out throughout the world have been targeted at establishing protected areas (Pressey, 1996). These areas are defined by the World Conservation Union as 'an area of land and/or sea especially dedicated to the protection and maintenance of biological diversity, and of natural and associated cultural resources, and managed through legal or other effective means' (IUCN, 1994). Protected areas act as refuges for species and ecological processes that cannot persist in intensely modified landscapes and seascapes and provide space for natural evolution and future ecological restoration (Chape et al., 2008; Gaston et al., 2008; Dudley et al., 2010). In many cases, they are the only remaining natural or semi-natural areas in whole regions, and significant numbers of species are found nowhere else (Ricketts et al., 2005). But protected areas do not just act as a refuge for biodiversity: their value also lies in the protection of cultural heritage (Fiske, 1992) and the provision of a range of socio-economic benefits (Ezebilo & Mattsson, 2010). Protected areas can supply a range of ecosystem services such as clean water, wild food, local medicines or genetic material; they can sometimes help maintain microclimatic or climatic stability; they can generate revenues through tourism and they can also help mitigating the impacts of climate change through carbon capture and storage (Chape et al., 2008; IPCC Secretariat, 2008; Dudley et al., 2010). The number and extent of protected areas have grown dramatically over recent decades (Coad et al., 2012), and approximately one-eighth of the Earth's terrestrial surface has now been formally designated as protected areas

Protected Areas: Are They Safeguarding Biodiversity?, First Edition.
Edited by Lucas N. Joppa, Jonathan E. M. Baillie and John G. Robinson.

(McDonald & Boucher, 2010), thus representing one of the most significant resource allocations on the planet (Balmford et al., 2003). Their static nature makes them particularly vulnerable to global environmental change, yet at the same time, their role as indicators of ecosystem responses to global environmental change is gaining support. Because of their importance and relative vulnerability, monitoring their status and performance is essential for making relevant management decisions in the face of future environmental change (Gaston et al., 2008; Fuller et al., 2010).

Satellite-based approaches have been highlighted by many as a cost-effective method to support monitoring efforts of protected areas, offering a cheap, verifiable way to identify areas of concern at a global scale and to support managers in their effort to design and apply adaptive management strategies (Gillespie et al., 2008; Alcaraz-Segura et al., 2009; Nemani et al., 2009; Pettorelli et al., 2012). There are many ways in which satellite data can inform protected area monitoring, such as helping monitor changes in land cover and land use, in ecosystem functioning and in the level of various human induced disturbances in and around protected areas, or providing information on processes and phenomena that act over larger areas and shape local environmental conditions experienced by wildlife. In this chapter, we will first highlight the range of opportunities existing for space-borne environmental information to support monitoring efforts of terrestrial protected area worldwide. We will then focus on existing, global products that are of value for monitoring protected areas, as shown by the Digital Observatory for Protected Areas (DOPA) (DOPA, 2013), an initiative aiming to assess, monitor and forecast the state and pressure of protected areas at the global scale (Dubois et al., 2011). DOPA is built around a set of distributed databases and modelling web-based services developed at different institutions like the Joint Research Centre (JRC) of the European Commission, the Global Biodiversity Information Facility (GBIF), the UNEP World Conservation Monitoring Centre (WCMC), Birdlife International and the International Union for Conservation of Nature (IUCN). It also is a contribution to the Group on Earth Observations Biodiversity Observation Network (GEO BON), the biodiversity arm of the Global Earth Observation System of Systems. In the second part of this chapter, we will draw attention to the opportunities existing for satellite-based information to support the monitoring of marine protected areas. This section, among other things, will be informed by recent work done by the JRC and the International Council for the Exploration of the Sea, who has been tasked to develop and implement cost-effective measures necessary to achieve or maintain good environmental status in the marine environment. In the third part of this chapter, we will discuss future opportunities for satellite-based data to support protected area management in the terrestrial and oceanic realms, highlighting some of the new technological developments and recent research that could orientate future efforts to further develop remote sensing-based monitoring tools. The last section of this chapter will discuss current and future limitations, which need to be addressed to realistically implement the effective and long-term use of satellite data to support protected area management worldwide.

Remote sensing and terrestrial protected area monitoring

Monitoring changes in ecosystems distribution

Characterising ecosystems, monitoring its land cover status and changes in their spatial arrangement, which is an indispensable process for assessing connectivity between protected areas, is a complex task that can be greatly facilitated by remote sensing products. Satellites collect a range of information that can be used to derive parameters which help to characterise ecosystems, ranging from bioclimatic variables (such as temperature, evapotranspiration or rainfall) to topographic variables (elevation), land cover (classifications or percentage of land cover types) and vegetation indices. By mixing such remotely sensed information, one can compute over a whole ecoregion the probabilities to find a combination of these variables that is similar to the combination found in the selected protected area of reference. The approach, traditionally used for ecological niche modelling, can be used to characterise the level of uniqueness of a protected area as well as to assess its isolation with respect to areas with similar environmental conditions in its surrounding (Rotenberry et al., 2006; Hartley et al., 2007; Dubois et al., 2013).

Monitoring changes in ecosystem functions

Primary productivity is the most integrative indicator of ecosystem functioning (Monteith, 1981) and a major 'supporting' ecosystem service, required for the production of many other ecosystem services (MEA, 2005). This parameter can be approximated using the Normalised Difference Vegetation Index (NDVI) (calculated from spectral reflectance measurements in the red and near-infrared regions), which is a remote sensing-based variable linearly related to the fraction of photosynthetically active radiation absorbed by the vegetation (Pettorelli et al., 2005). Radiation interception is the main control of carbon gains and thus of net primary productivity. NDVI data can be derived from various multispectral sensors and satellites (Landsat, AVHRR, MODIS, SPOT), and several studies have recently highlighted how spatiotemporal fluctuations in this vegetation index can inform protected area management worldwide (Alcaraz-Segura et al., 2009; Tang et al., 2011; Pettorelli et al., 2012). Over the last decades, the NDVI has also proven extremely useful in predicting animal distribution and the abundance and life history traits of species in space and time (Pettorelli, 2013).

Monitoring anomalies and disturbances

Climatic anomalies

Climatic conditions can strongly impact the functioning of ecosystems. Climatic anomalies that can potentially impact protected areas can be detected and monitored by contrasting current climatic records with seasonal norms of rainfall or temperature. As proposed by Hartley et al. (2007) in their African Protected Areas Assessment Tool, these anomalies can be characterised by their strength, duration and deviation from their expected occurrence in time (Figure 14.1). The strength here represents how different the current record is from the expected value: the bigger the difference, the greater the strength. The duration is simply the length of the period over which the anomaly has been observed. The deviation is an index that indicates whether the current record actually is an early or late event (such as an early start to the rainy season). Several sources of satellite rainfall estimates can be used to monitor rainfall anomalies: examples of sources for rainfall estimates include FEWS NET from the USGS, TAMSAT from the University of Reading and the near real-time 15 minutes Multi-Sensor Precipitation Estimate from EUMETSAT.

Fire

Fire is a crucial ecological component of many ecosystems. Fire activity in protected areas can be monitored using MODIS fire products available through the National Aeronautics and Space Administration (NASA)-funded Fire Information for Resource Management System. Information on fire occurrence is derived four times a day by the MODIS sensors onboard the TERRA and AQUA satellites. Active fires data are available with a 1 km spatial resolution, while the information on the burned areas extent is provided at 500 m resolution. These data can be used not only to characterise the spatial and temporal distribution of fires but also to define the fire intensity using the fire radiative power provided by the MODIS active fire product. The correlation between fire intensity and the fuel load can be further used to assess land cover change and landscape fragmentation around protected areas (Palumbo et al., 2011).

Remote sensing and marine protected area monitoring

Monitoring changes in ecosystems distribution

While over 13% of the terrestrial environment is under protection (Chape et al., 2008; Coad et al., 2012), recent calculations by the Marine Reserve Coalition show that only 3.2% of the marine environment is protected (MRC, 2012) (the proportion of this that is effectively so is likely to be much lower). This figure falls short of the 10% coverage target meant to be achieved by 2012, which was adopted by the Convention on Biological Diversity in 2006 (Annex IV, Target 1.1) (CBD, 2006). The 2012 target date was extended by 2020 after the failure to attain 10% by the original target date (Strategic Goal C, Target 11) (CBD, 2010), although some called for increasing the target percentage up to 30% (Sheppard et al., 2012). In addition, Target 11 specifies that the 10% must be regionally representative of ecoregions and habitats. Bio-regionalisation is important for policy, yet for many

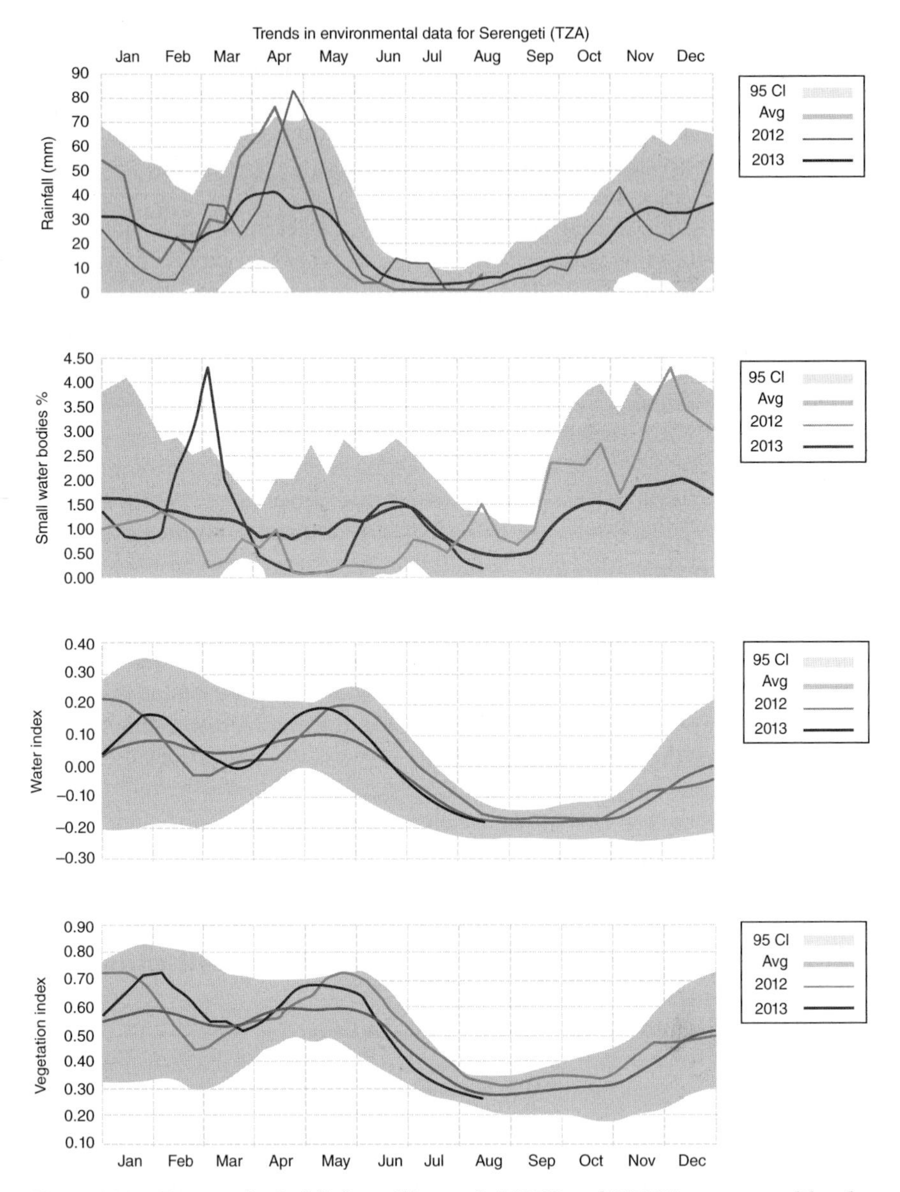

Figure 14.1 **Excess of rainfall (in millimetre), NDVI and NDWI as reported by the African Protected Areas Assessment Tool (Hartley et al., 2007) for the Serengeti National Park (Tanzania). Observations are collected for each 10-day period and contrasted against 10-year averages (grey line) and associated 95% Confidence intervals (grey areas). Source: Data from Hartley et al. (2007)**

marine habitats, their extension is not yet fully established. This makes accomplishing the target conservation status more challenging. Open water habitats in particular lag behind in terms of their protected area status with 1.79% of waters deeper than 200 m protected, in comparison to 7.9% of continental slope and similar areas (<200 m deep) (Spalding et al., 2013). The combination of high spatial and temporal coverage of satellite remote sensing data has the potential to help characterise marine environments. This would help identify relevant habitats for protection and subsequent monitoring, even if information below the sea surface will remain very difficult to collect (Kachelriess et al., 2014).

As demonstrated for terrestrial protected areas, a range of parameters routinely collected by satellite can be used to characterise the marine environment and support marine protected areas management (Kachelriess et al., 2014). These include physical parameters such as sea surface temperature (SST, °C), sea surface salinity (SSS, psu), sea surface height (SSH, cm) and wind speed and direction (m/s). In addition, a measure of biological activity can also be attained through ocean colour data (cholorophyll-a), and with additional parameters including photosynthetically available radiation (PAR), primary productivity can be estimated (Melin & Hoepffner, 2011). Such variables used individually (Esaias et al., 2000) or in combination (Devred et al., 2007; Dowell & Platt, 2009), aggregated monthly, seasonally or annually (Gregr & Bodtker, 2007), or using climatological means (Hardman-Mountford et al., 2008; Vichi et al., 2011), can be used to delineate ecologically meaningful classification systems of the pelagic environment. Carr et al. (2011) identified key oceanographic variables that need to be monitored to assess ecosystem change and marine protected area effectiveness, along with additional parameters such as secondary production (including zooplankton diversity and biomass), nutrients and bottom/habitat characteristics (Carr et al., 2011).

These variables can be used to compute within an ecoregion the probability of finding a combination of such variables that is similar to the combination found in the selected protected area. This allows for the assessment of the degree of similarity between habitats and assessment of the connectivity between habitats. An important consideration when applying such a tool to the marine environment is the dynamic nature of the oceans (both spatially and temporally) as well as the vertical heterogeneity and tools need to be developed that characterise this nature. When effectively developed, such tools can be used to monitor changes in environmental variables that result in alterations in habitats to allow for scenario testing of the potential effects of climate change.

Monitoring changes in ecosystem functions

As with terrestrial systems, primary productivity (mg C m^{-2} d^{-1}) is an integrator of ecosystem function. Remote sensing has greatly increased the spatial and temporal extent of primary production data as compared to in situ measurements, and the use of such data has concurrently expanded to include a wide variety of applications. The detection of mesoscale oceanic features such as fronts, eddies and filaments can be made using satellite remote sensing data (Chassot et al., 2011). Such features are often associated with increased productivity and can help identify ecologically or biologically significant areas (EBSAs), such as feeding or spawning areas where species are seen to aggregate (Drunon, 2010; Drunon et al., 2011). The satellite-based identification of EBSAs has been effectively used by the fishing community to exploit marine resources with a significant increase in catch and associated decline in effort (Santos, 2000), leading to increased fisheries productivity (Mugo et al., 2010; Tseng et al., 2010). Interestingly, the spatial distribution of EBSAs has also been shown to match the distribution of preferred habitats for exploited species (Royer et al., 2004). Not only have EBSAs been linked to fish abundance but also to higher densities of all trophic levels of marine fauna, from zooplankton through to vertebrates, including marine mammals (e.g. whales, seals), seabirds (e.g. penguins, albatross) and reptiles (e.g. turtles).

Quantifying primary production required to sustain species of interest within marine systems has been used as an indicator of the capacity of systems to support populations. Estimates are often provided relative to the total net primary production of the system whether the target species are related to fisheries (Pauly & Christensen, 1995) or conservation (e.g. cetaceans) (Barlow et al., 2008). Improved estimates of net primary production from remote sensing data have allowed such indicators to be produced routinely when ecosystem level assessments are being made. Estimates of primary production has been used to look at what is thought to be bottom-up control in both the European Union seas (Chassot et al., 2007) as well as at a global scale (Chassot et al., 2010). Changes in primary productivity, as determined from remotely sensed data in combination with future climate change models, have also been used to investigate a range of ecosystem indicators to determine the effect of such changes on systems (e.g. Australian ecosystems) (Brown et al., 2005).

Monitoring climatic anomalies

The generation of climatic anomalies (deviations from long-term means) to investigate changes in physical (e.g. sea surface temperature) and biological (e.g. chlorophyll-a) parameters through time has provided indicators that are useful for monitoring of marine systems at global, regional and local levels. Temperature is an important variable that delineates tolerance envelopes for marine species. Remotely sensed temperature anomalies have been linked to both coral bleaching and disease (Selig et al., 2010), and considering such anomalies might be relevant for fulfilling conservation objectives on the Great Barrier Reef (Ban et al., 2012). While the association of species distribution to areas of elevated chlorophyll-a or primary production has been documented, chlorophyll-a anomalies have also been used to identify hotspots (e.g. for seabirds) (Suryan & Hay, 2012). The timing and duration of phytoplankton blooms detected from remotely sensed satellite data have been related to the recruitment success of target species (e.g. shrimps (Fuentes-Yaco et al., 2007; Koeller et al., 2009) and haddock (Platt et al., 2003, 2007)). A study which developed an index of larval retention used altimetry data to drive a larval transport model which could be useful for the same purpose (Polovina & Howell, 2005). Changes in oceanographic patterns have also been identified using remotely sensed data which has shown increased advection associated with El Niño events and related these to an increases in fisheries catch (e.g. bigeye tuna) (Howell & Kobayashi, 2006).

Future opportunities

Up to now, this chapter has detailed the main type of satellite-based information that is routinely used by global monitoring systems such as DOPA. However, there is a broader use of remote sensing imagery, in particular of high- and very high-resolution imagery which can further contribute to improve management of protected areas. Very high-resolution imagery can, for example, allow for some species identification (Carleer & Wolff, 2004), but the automatic and regular processing of such information for global assessments is still a computational challenge that will require significant innovation in the handling of large data sets as well as mapping algorithms. The following texts are some examples of these future avenues for protected area monitoring from space.

Monitoring animal distribution from space

The use of aerial images to count animals has become relatively common over the past two decades (Jachmann, 2002; Laliberte & Ripple, 2003; Descamps et al., 2011; Groom et al., 2011; Sirmacek et al., 2012). Counting animals using information collected by satellites is much less common, although examples do exist. One such example comes from the Antarctic, where high- and very high-resolution images have been used to detect and estimate emperor penguin colony sizes (Barber-Meyer et al., 2007; Fretwell & Trathan, 2009; Fretwell et al., 2012). More precisely, Landsat ETM images were used to detect faecal staining of ice by penguins and derive pan-Antarctic maps of locations of colonies (Fretwell & Trathan, 2009). Such information was then used to assemble relevant very high-resolution imagery from the QuickBird, WorldView-2 and Ikonos satellites and determine, for each location, the colony size (Fretwell et al., 2012). In this particular situation, the use of remote sensing data dramatically enhanced our current knowledge of this species' distribution and provided, for the first time, a technique for long-term monitoring across its distributional range. Yet, as highlighted by the authors, many other species could benefit by such an approach – although in its infancy, the direct monitoring of animals from space is a promising avenue to explore and has the potential to inform protected area management in many open areas, starting with the Antarctic.

Detecting and monitoring invasive species

Invasive alien species can have many adverse impacts on biodiversity, ecosystem services, agriculture, forestry, the economy and human welfare (MEA, 2005). Possible invasions by such species therefore represent a real threat to protected areas management. Remote sensing data can be used for the detection and monitoring of invasive alien species. As previously mentioned, maps of habitat similarities can also be used to assess and possibly forecast dispersion of alien invasive species. For example, a recent paper by Dlamini (2012) describes the use of probabilistic graphical models, in particular Bayesian networks, for the detection of two invasive alien plants, *Chromolaena odorata* and *Lantana camara*, from the 8-band Worldview-2 satellite imagery over central Swaziland (Dlamini, 2012). Using the spectral information collected by Landsat 5 TM and Landsat 7 ETM+, Wilfong et al. (2009) showed that the NDVI could be used to map current and past understory invasion by Amur honeysuckle (*Lonicera maackii*) (Wilfong et al., 2009). Venugopal (1998) used this same vegetation index to monitor the infestation of water hyacinths (*Eichhornia crassipes*) in Bangalore (Venugopal, 1998), while more recently, Hoyos et al. (2010) used the NDVI to map glossy privet (*Ligustrum lucidum*) invasion in Argentina (Hoyos et al., 2010). Hyperspectral data seem to be particularly effective at mapping the distribution of invasive species (He et al., 2011). This was well illustrated by the work by Asner and colleagues, who showed that the hyperspectral signature of Hawaiian native trees was generally unique from those of introduced trees (Asner et al., 2008). Information collected by these hyperspectral sensors moreover allowed for monitoring the impact of the invasion at the ecosystem level, by providing data on biochemical changes found at the foliar and canopy levels. Future space-borne sensors by spatial agencies such as NASA and the German Aerospace Centre (DLR) will hopefully provide a global coverage on more appropriate spatial and temporal scales.

Monitoring threats to terrestrial protected areas: The case of encroachment and land cover change

The regular monitoring of land cover in and around protected areas can be used to map human disturbances permeating into the parks. Assessments of land cover change around protected areas can be used to measure increasing pressures on protected areas as well as the potential loss of connectivity of the park with its neighbouring natural ecosystems. Similarly, observations made inside protected areas can further be used to detect not only encroachments and potential loss of habitats but also other potential sources of conservation conflicts. The detection of land cover changes in habitats like croplands, grasslands, savannahs, shrub lands and mountain areas as well as in heterogeneous regions however remains very much a challenge as encroachments are difficult to spot, while the automation of the detection of changes is still in its infancy (Gross et al., 2013). Significant efforts are devoted to put in place a reliable global monitoring framework for land cover change in protected areas, and this feature is expected to be part of DOPA in the years to come.

Monitoring threats to marine protected areas: The case of oil spill, illegal fishing and run-off

Oil spills can cause significant short-term and long-term damages to marine biodiversity. Detecting the emergence of such a threat, and monitoring its progression, could be key for mitigating its impact on marine protected areas. Various established remote sensing-based methods exist to detect oil spills and track their development, for example, using Synthetic Aperture Radar (SAR) or infrared sensors (Klemas, 2010; Ottavaini et al., 2012). Hyperspectral data can also be used to discriminate hydrocarbons and track oil spills (Horig et al., 2001). Illegal, undeclared or unreported fishing is another major threat to the successful implementation of marine protected areas (Game et al., 2009). To enforce spatial closures and other fisheries management policies, most nations require fishing vessels to carry a vessel monitoring system. To elude the regulation, however, fishing vessels can turn off their vessel monitoring system. Combining remote sensing information with data from active vessel monitoring systems can allow ships that have switched their systems off to be detected, as high-resolution optical satellites can be used to detect vessels and monitor vessel movement (Corbane et al., 2008, 2010). SAR data sets are, besides passive remote sensing, another promising approach to detect ships (Kachelriess et al., 2014).

Limitations

Remote sensing cannot solve the entire information collection issue associated with protected area monitoring. Remote sensing-based techniques can address spatial and temporal domains inaccessible to traditional, on the ground, approaches but cannot match the accuracy, precision and thematic richness of in situ measurement and monitoring (Gross et al., 2009). For example, satellite-based information mainly provides information on changes in canopy cover in terrestrial forested areas; yet biodiversity loss is known to happen in perfectly healthy forests, such as the de-faunation associated with the empty forest syndrome (Redford, 1992; Bennett et al., 2002; Wilkie et al., 2011). This example illustrates the important point that remote sensing-based monitoring methods need to be integrated with field observations to maximise monitoring effectiveness.

Currently available spatial and temporal resolutions can sometimes be viewed as a technical constraint, limiting opportunities for small-scale habitat classification (Purkis & Klemas, 2011). Available spectral resolution can also restrict the usefulness of satellite data for protected area management, as it can be difficult to monitor specific plant species (e.g. many invasive plant species) without hyperspectral sensors (He et al., 2011). Importantly, average level of cloud cover or atmospheric disturbances, average above ground biomass density, available financial and human resources, protected area size, sought accuracy and resolutions are all factors that are spatially and temporally variable and can affect the usefulness of remote sensing data. Therefore, efforts to set up a remote sensing-based framework for protected area management should be focused on developing relevant frameworks at multiple spatial scales (i.e. global, regional, national and local).

For a realistic implementation of satellite data to support protected area management, financial and human resources need to be taken into account. While a lot of satellite data is now freely available, the very high-resolution imagery necessary for several of the applications described in this chapter need to be purchased (e.g. WorldView-2, IKONOS). The comparatively large volumes of data (e.g. above 100 MB for one Landsat-5 TM scene 185 km (185 km with 30 m resolution) in GEOTIFF format) also potentially pose a logistical challenge for data transmission, storage and processing of large areas and time series. Another cost factor is associated with the software and expertise required to process and analyse remote sensing data collected at different spatial, temporal and spectral resolutions: while open source solutions exist (Steiniger & Hay, 2009; Rocchini & Neteler, 2012), commercial software solutions still dominate the education and training of the workforce. Remote sensing-based monitoring of protected areas is also not a 'one-off' process, and it will be therefore essential to have guaranteed access to data, expertise and software in the long term. This means that local stakeholders, governments and spatial agencies all need to commit to a data policy that allows remote sensing-based monitoring. Altogether, such costs and commitments can be perceived as being considerable for some of the countries where remote sensing monitoring of the protected area network would be the most beneficial. The methods, cost and benefits of traditional on the ground data collection protocols are moreover relatively familiar to environmental decision makers, as opposed to the complexity associated with the various components of remote sensing, which may also preclude its adoption (Strand et al., 2007).

Conclusions

The aim of this chapter was to illustrate the current and future potential for remote sensing information to guide protected area designation and management. From this brief review and series of examples, it is evident that the potential is there (Nagendra et al., 2013). So why are remote sensing-based monitoring frameworks for protected areas still in their infancy?

One reason is the difficulty of making the imagery and associated processing tools available in developing, and often biodiversity-rich, countries. Imagery is difficult to exchange over the Internet and access to broadband Internet is clearly a huge obstacle for many users in developing countries. In contrast to most information processed by DOPA that is exchanged over the Internet, its remote sensing module (eStation) receives the data mainly through EUMETCast, a satellite-based data dissemination system, providing DOPA with some means to be partly independent from an Internet-based infrastructure (Clerici et al., 2013). The eStation, initially developed as an independent module processing service, is already used in around 50 African countries, and a similar infrastructure could be more widely used in areas lacking sufficient access to the Internet. A second reason is the limited dialogue between the remote sensing community and the community of conservationists and environmental managers: there is indeed a dramatic need to build bridges between those processing the imagery and those actively applying such data for conservation purposes. A third reason is the key importance that conservationists and environmental managers place on being able to monitor land cover change, something that has proven difficult to achieve using remote sensing information.

Frustrations associated with the difficulties of accessing this important information might perhaps have led to a disinterest in remote sensing for many conservationists and managers. Similarly, frustrations associated with the difficulties of accessing reliable, standardised, on the ground information to validate remote sensing products or to assess their ecological relevance have potentially led to a lack of interest in conservation and environmental management for many remote sensing experts.

So how do we move forward? There are several opportunities that could facilitate the use of remote sensing products for conservation purposes and therefore pave the way for developing an enhanced remote sensing-based monitoring framework for protected areas. First, better and stronger communication channels need to be established between the remote sensing community, conservationists and environmental managers to start developing a coordinated, effective research agenda. Recent efforts towards the establishment of such bridges are encouraging, as there is a perceptible and growing interest in and need for remote sensing applications within the conservation community. This is well illustrated by recent initiatives to identify remote sensing requirements for biodiversity research and conservation applications, which include a DLR-funded Committee on Earth Observation Satellites (CEOS) workshop in Munich (October 2012) and a NASA-funded workshop to identify the top 10 conservation questions that can be addressed using remote sensing in Washington (January 2013). This is also exemplified by the large attendance of conservation remote sensing discussions at the International Conference on Conservation Biology (ICCB) in 2011, 2013 and 2015 and at the Society for Conservation GIS (SCGIS) 2012 conference in Pacific Grove. Platforms facilitating exchanges of information and networking opportunities between communities are also flourishing, such as the remote sensing and conservation website (RSEC, 2012). GEO BON is working to identify the Essential Biodiversity Variables to be focusing on in the future, as well as working towards identifying gaps in current earth observation products. Such an international dialogue on how and what should be monitored is bound to help research communities develop common semantic and identify obvious synergies. Various citizen science projects could also help refine products and strengthen the link between different communities, as well as help build databases on wildlife distribution. Finally, open access databases (such as GBIF or Movebank) and liberal data policies (such as free Landsat archive) and open source solutions are advancing, which makes access and data/knowledge exchange easier.

A strong commitment to focus on the development of practical applications of satellite remote sensing for the monitoring of protected areas would significantly improve the outcomes of conservation efforts, at both local and global scales. Much remains to be done to first ensure the setting up of well-orchestrated biodiversity informatics infrastructures capable of supporting the needs of both communities. A mutual willingness to collaborate and develop joint approaches has already emerged and is capable to act as a solid foundation for the necessary future endeavours.

References

Alcaraz-Segura, D., Cabello, J., Paruelo, J. M. & Delibes, M., 2009. Use of descriptors of ecosystem functioning for monitoring a national park network: A remote sensing approach. *Environmental Management*, Volume 43, p. 38–48.

Asner, G. P. et al., 2008. Remote sensing of native and invasive species in Hawaiian forests. *Remote Sensing and Environment*, Volume 112, p. 1912–1926.

Balmford, A. et al., 2003. Global variation in terrestrial conservation costs, conservation benefits and unmet conservation needs. *Proceedings of the National Academy of Sciences of the United States of America*, Volume 100, p. 1046–1050.

Ban, N. C., Pressey, R. L. & Weeks, S., 2012. Conservation objectives and sea surface temperature anomalies in the Great Barrier Reef. *Conservation Biology*, Volume 26, p. 799–809.

Barber-Meyer, S. M., Kooyman, G. L. & Ponganis, P. J., 2007. Estimating the relative abundance of emperor penguins at inaccessible colonies using satellite imagery. *Polar Biology*, Volume 30, p. 1565–1570.

Barlow, J., Kahru, M. & Mitchell, B. G., 2008. Cetacean biomass, prey consumption and primary production requirements in the Californian Current Ecosystem. *Marine Ecology Progress Series*, Volume 371, p. 385–295.

Bennett, E., Eves, H., Robinson, J. & Wilkie, D., 2002. Why is eating bushmeat a biodiversity crisis. *Conservation Biology in Practice*, Volume 3(1), p. 28–29.

Brown, C. W. et al., 2005. An introduction to satellite sensors, observations and techniques. In: R. L. Miller, C. E. Del Castillo & B. A. McKee, eds. *Remote sensing of coastal aquatic environments – Technologies, techniques and applications*. Dordrecht: Springer, p. 21–50.

Carleer, A. & Wolff, E., 2004. Exploitation of very high resolution satellite data for tree species identification. *Photogrammetric Engineering and Remote Sensing*, Volume 70, p. 135–140.

Carr, C. M. et al., 2011. A tri ocean perspective: DNA barcoding reveals geographic structure and cryptic diversity in Canadian polychaetes. *PLoS ONE*, 6(7), p. e22232.

CBD, 2006. *Decisions adopted by the CoP to the CBD at its 8th meeting (UNEP/CBD/COP/DEC/VIII/15)*, Montreal: Secretariat of the CBD.

CBD, 2010. *Decisions adopted by the CoP to the CBD at its 10th meeting (UNEP/CBD/COP/DEC/X/2)*, Montreal: Secretariat of the CBD.

Chape, S., Spalding, M. & Jenkins, M., 2008. *The world's protected areas: Status, values and prospects in the 21st century*, Berkeley: University of California Press.

Chassot, E., Melin, F., Le Pape, O. & Gascuel, D., 2007. Bottom up control regulates fisheries production at the scale of eco regions in European seas. *Marine Ecology Progress Series*, Volume 343, p. 45–55.

Chassot, E. et al., 2010. Global marine primary production constrains fisheries catches. *Ecology Letters*, Volume 13, p. 495–505.

Chassot, E. et al., 2011. Satellite remote sensing for an ecosystem approach to fisheries management. *ICES Journal of Marine Science*, Volume 68, p. 651–666.

Clerici, M. et al., 2013. The eStation, an earth observation processing service in support to ecological monitoring. *Ecological Informatics*, Volume 18, p. 162–170.

Coad, L., Burgess, N. D., Bombard, B. & Besancon, C., 2012. *Progress towards the CBD's 2010 and 2012 targets for protected area coverage*, Cambridge: UNEP-WCMC.

Corbane, C., Marre, F. & Petit, M., 2008. Using SPOT-5 HRG data in panchromatic mode for operational detection of small ships in tropical areas. *Sensors*, Volume 8, p. 2959–2973.

Corbane, C. et al., 2010. A complete processing chain for ship detection using optical satellite imagery. *International Journal of Remote Sensing*, Volume 31, p. 5837–5854.

Descamps, S. et al., 2011. An automatic counter for aerial images of aggregations of large birds. *Bird Study*, Volume 58, p. 302–308.

Devred, E., Sathyendranath, S. & Platt, T., 2007. Delination of ecological provinces using ocean colour radiometry. *Marine Ecology Progress Series*, Volume 346, p. 1–13.

Dlamini, W. M., 2012. Exelis: Visual Information Solutions. [Online] Available at: http://www.exelisvis.com/portals/0/pdfs/envi/8_bands_Wisdom_Dlamini.pdf [Accessed 18 December 2012].

DOPA, 2013. Digital Observatory for Protected Areas. [Online] Available at: http://dopa.jrc.ec.europa.eu [Accessed 7 August 2013].

Dowell, M. & Platt, T., 2009. *Partition of the ocean into ecological provinces: Roll of ocean colour radiometry*, Dartmouth: IOCCG.

Drunon, J. N., 2010. Habitat mapping of the Atlantic bluefin tuna derived from satellite data: Its potential as a tool for the sustainable management of pelagic fisheries. *Marine Policy*, Volume 34, p. 293–297.

Drunon, J. N., Fromentin, J. M., Aulanier, F. & Heikkonen, J., 2011. Potential feeding and spawning habitats of bluefin tuna in the Mediterranean Sea. *Marine Ecology Progress Series*, Volume 439, p. 223–240.

Dubois, G. et al., 2011. *Proceedings of the 34th international symposium on remote sensing of environment*, Sydney: ICRSE.

Dubois, G. et al., 2013. eHabitat, a multipurpose web processing service for ecological modelling. *Environmental Modelling and Software*, Volume 41, p. 123–133.

Dudley, N. et al., 2010. *Natural solutions: Protected areas helping people cope with climate change*, Gland, Washington, DC & New York: IUCN-WCPA, TNC, UNDP, WCS, World Bank & WWF.

Esaias, W. E., Iverson, R. L. & Turpie, K., 2000. Ocean province classification using ocean colour data: Observing biological signatures of variations in physical dynamics. *Global Change Biology*, Volume 6, p. 39–55.

Ezebilo, E. E. & Mattsson, L., 2010. Socioeconomic benefits of protected areas as perceived by local people around Cross River National Park, Nigeria. *Forest Policy and Economics*, Volume 12, p. 189–193.

Fiske, S. J., 1992. Sociocultural aspects of establishing marine protected areas. *Ocean and Coastal Management*, Volume 18, p. 25–46.

Fretwell, P. T. & Trathan, P. N., 2009. Penguins from space: Faecal stains reveal the location of emperor penguin colonies. *Global Ecology and Biogeography*, Volume 18, p. 543–552.

Fretwell, P. T. et al., 2012. An emperor penguin population estimate: The first global synoptic survey of a species from space. *PLoS ONE*, Volume 7(4), e33751.

Fuentes-Yaco, C., Koeller, P. A., Sathyendranath, S. & Platt, T., 2007. Shrimp (*Pandalus borealis*) growth and timing of the spring phytoplankton bloom on the Newfoundland-Labrador Shelf. *Fish Oceanography*, Volume 16, p. 116–129.

Fuller, R. A. et al., 2010. Replacing underperforming protected areas achieves better conservation outcomes. *Nature*, Volume 466, p. 365–367.

Game, E. T. et al., 2009. Pelagic protected areas: The missing dimension in ocean conservation. *Trends in Ecology & Evolution*, Volume 24, p. 360–369.

Gaston, K. J., Jackson, S. F., Cantu-Salazar, L. & Cruz-Pinon, G., 2008. The ecological performance of protected areas. *Annual Review of Ecology, Evolution and Systematics*, Volume 39, p. 99–113.

Gillespie, T. W. et al., 2008. Measuring and modelling biodiversity from space. *Progress in Physical Geography*, Volume 32(2), p. 203–221.

Gregr, E. J. & Bodtker, K. M., 2007. Adaptive classification of marine ecosystems: Identifying biologically meaning regions in the marine environment. *Deep-Sea Research*, Volume 54, p. 385–402.

Groom, G., Petersen, I. K., Anderson, M. D. & Fox, A. D., 2011. Using object based analysis of image data to count birds: Mapping of lesser flamingos at Kamfers dam Northern Cape, South Africa. *International Journal of Remote Sensing*, Volume 32, p. 4611–4639.

Gross, J. E., Goetz, S. J. & Cihlar, J., 2009. Application of remote sensing to parks and protected area monitoring: Introduction to the special issue. *Remote Sensing of Environment*, Volume 113, p. 1343–1345.

Gross, D. et al., 2013. Monitoring land cover changes in African protected areas in the 21st century. *Ecological Informatics*, Volume 14, p. 31–37.

Hardman-Mountford, N. J., Hirata, T., Richardson, K. A. & Aiken, J., 2008. An objective methodology for the classification of ecological pattern into biomes and provinces for the pelagic ocean. *Earth Observations for Marine and Coastal Biodiversity and Ecosystems Special Issue*, Volume 112(8), p. 3341–3352.

Hartley, A., Nelson, A., Mayaux, P. & Gregoire, J.-M., 2007. *The assessment of African protected areas – JRC scientific and technical reports*, Luxembourg: Office for Official Publications of the European Communities.

He, K. S., Rocchini, D., Neteler, M. & Nagendra, H., 2011. Benefits of hyperspectral remote sensing for tracking plant invasions. *Diversity and Distributions*, Volume 17, p. 381–392.

Horig, B., Kuhn, F., Oschutz, F. & Lehmann, F., 2001. HyMap hyperspectral remote sensing to detect hydrocarbons. *International Journal of Remote Sensing*, Volume 22, p. 1413–1422.

Howell, E. A. & Kobayashi, D. R., 2006. El Nino effects in the Palmyra Atoll region: Oceanographic changes and bigeye tuna (*Thunnus obesus*) catch rate variability. *Fisheries Oceanography*, Volume 15, p. 477–489.

Hoyos, L. E. et al., 2010. Invasion of glossy privet (*Ligustrum lucidum*) and native forest loss in the Sierras Chicas of Cordoba, Argentina. *Biological Invasions*, Volume 12, p. 3261–3275.

IPCC Secretariat, 2008. *Fourth assessment report of the Intergovernmental Panel on Climate Change*, Geneva: IPCC.

IUCN, 1994. *Guidelines for protected area management categories*, Gland & Cambridge: IUCN.

Jachmann, H., 2002. Comparison of aerial counts with ground counts for large African herbivores. *Journal of Applied Ecology*, Volume 39, p. 841–852.

Kachelriess, D., Wegmann, M., Gollock, M. & Pettorelli, N., 2014. The application of remote sensing for marine protected area management. *Ecological Indicators*, Volume 36, p. 169–177.

Klemas, V., 2010. Tracking oil slicks and predicting their trajectories using remote sensors and models: Case studies of the Sea Princess and Deepwater Horizon oil spills. *Journal of Coastal Research*, Volume 265, p. 789–797.

Koeller, P. et al., 2009. Basin scale coherence in phenology of shrimps and phytoplankton in the North Atlantic Ocean. *Science*, Volume 324, p. 791–793.

Laliberte, A. S. & Ripple, W. J., 2003. Automated wildlife counts form remotely sensed imagery. *Wildlife Society Bulletin*, Volume 31, p. 362–371.

McDonald, R. I. & Boucher, T. M., 2010. Global development and the future of the protected area strategy. *Biological Conservation*, Volume 144, p. 383–392.

MEA, 2005. *Ecosystems and human well being: Biodiversity Synthesis*, Washington, DC: World Resources Institute.

Melin, F. & Hoepffner, N., 2011. Case Study 6 – Monitoring phytoplankton productivity from satellite: An aid to marine resources management. In: J. Morales, V. Stuart, T. Platt & S. Sathyendranath, eds. *Handbook of satellite remote sensing image interpretation: Applications for marine living resources conservation and management*. Dartmouth: EU PRESPO & IOCCG, p. 79–93.

Monteith, J. L., 1981. Climatic variation and the growth of crops. *Quarterly Journal of the Royal Meteorological Society*, Volume 107, p. 749–774.

MRC, 2012. The Race to Protect Our Oceans – MRC Delivers MPA League Table at Rio +20. [Online] Available at: http://www.marinereservescoalition.org/2012/06/18/the-race-to-protect-our-oceans-mrc-delivers-mpa-league-table-at-rio-20/ [Accessed 7 August 2013].

Mugo, R., Saitoh, S., Nihira, A. & Kuroyama, T., 2010. Habitat characteristics of skipjack tuna (*Katsuwonus pelamis*) in the western North Pacific: A remote sensing perspective. *Fish Oceanography*, Volume 19, p. 382–396.

Nagendra, H. et al., 2013. Remote sensing for conservation monitoring: Assessing protected areas, habitat extent, habitat condition, species diversity and threats. *Ecological Indicators*, Volume 33, p. 45–59.

Nemani, R. et al., 2009. Monitoring and forecasting ecosystem dynamics using the terrestrial observation and prediction system (TOPS). *Remote Sensing of the Environment*, Volume 113, p. 1497–1509.

Ottavaini, M. et al., 2012. Polarimetric retrievals of surface and cirrus cloud properties in the region affected by the Deepwater Horizon oil spill. *Remote Sensing of the Environment*, Volume 121, p. 389–403.

Palumbo, I., Gregoire, J. M., Simonetti, D. & Punga, M., 2011. Spatio temporal distribution of fire activity in protected areas of Sub-Saharan African derived from MODIS data. *Procedia Environmental Sciences*, Volume 7, p. 26–31.

Pauly, D. & Christensen, V., 1995. Primary production required to sustain global fisheries. *Nature*, Volume 374, p. 255–257.

Pettorelli, N., 2013. *The normalized differential vegetation index*, Oxford: Oxford University Press.

Pettorelli, N. et al., 2005. Using the satellite derived NDVI to assess ecological responses to environmental change. *Trends in Ecology & Evolution*, Volume 20, p. 503–510.

Pettorelli, N. et al., 2012. Tracking the effect of climate change on ecosystem functioning using protected areas: Africa as a case study. *Ecological Indicators*, Volume 20, p. 269–276.

Platt, T., Fuentes-Yaco, C. & Frank, K. T., 2003. Spring algal bloom and larval fish survival. *Nature*, Volume 423, p. 398–399.

Platt, T., Sathyendranath, S. & Fuentes-Yaco, C., 2007. Biological oceanography and fisheries management: Perspective after 10 years. *ICES Journal of Marine Science*, Volume 64, p. 893–869.

Polovina, J. J. & Howell, E. A., 2005. Ecosystem indicators derived from satellite remotely sensed oceanographic data for the north Pacific. *ICES Journal of Marine Science*, Volume 62, p. 319–327.

Pressey, R. L., 1996. Protected areas: Where should they be and why should they be there. In: I. F. Spellerberg, ed. *Conservation biology*. Harlow: Longman, p. 171–185.

Purkis, S. J. & Klemas, V., 2011. *Remote sensing and global environmental change*, 1 ed., Chichester: John Wiley & Sons, Ltd.

Redford, K. H., 1992. The empty forest. *Bioscience*, Volume 42, p. 412–422.

Ricketts, T. H. et al., 2005. Pinpointing and preventing imminent extinctions. *Proceedings of the National Academy of Sciences of the United States of America*, 102(51), p. 18497–18501.

Rocchini, D. & Neteler, M., 2012. Let the four freedoms paradigm apply to ecology. *Trends in Ecology & Evolution*, Volume 27, p. 310–311.

Rotenberry, J. T., Preston, K. L. & Knick, S. T., 2006. GIS based niche modelling for mapping species' habitat. *Ecology*, Volume 87, p. 1458–1464.

Royer, F., Fromentin, J.-M. & Gaspar, P., 2004. The association between bluefin tuna schools and oceanic features in the Western Mediterranean Sea. *Marine Ecology Progress Series*, Volume 269, p. 249–263.

RSEC, 2012. Remote Sensing for Conservation and Biodiversity Research. [Online] Available at: http://www.remote-sensing-conservation.org [Accessed 18 December 2012].

Santos, A. M. P., 2000. Fisheries oceanography using satellite and airborne remote sensing methods: A review. *Fisheries Research*, Volume 49, p. 1–20.

Selig, E. R., Casey, K. S. & Bruno, J. B., 2010. New insights into global patterns of ocean temperature anomalies: Implications for coral reef health and management. *Global Ecology and Biogeography*, Volume 19, p. 397–411.

Sheppard, C. R. C. et al., 2012. Reefs and islands of the Chagos Archipelago, Indian Ocean: Why it is the world's largest no-take marine protected area. *Aquatic Conservation: Marine and Freshwater Ecosystems*, Volume 22(2), p. 232–261.

Sirmacek, B. et al., 2012. *Automatic population counts for improved wildlife management using aerial photography*, Leipzig: iEMSs.

Spalding, M. D. et al., 2013. Protecting marine spaces: Global targets and changing approaches. *Ocean Yearbook*, Volume 27, p. 213–248.

Steiniger, S. & Hay, G. J., 2009. Free and open source geographic information tools for landscape ecology. *Ecological Informatics*, Volume 4(4), p. 183–195.

Strand, H. et al., 2007. *Sourcebook on remote sensing and biodiversity indicators: CBD Technical Series 32*, Cambridge: UNEP-WCMC.

Suryan, S. & Hay, G. J., 2012. New approach for using remote sensed chlorophyll-a to indentify seabird hotspots. *Marine Ecology Progress Series*, Volume 451, p. 213–225.

Tang, Z., Fang, J., Sun, J. & Gaston, K. J., 2011. Effectiveness of protected areas in maintaining plant production. *PLoS ONE*, Volume 6(4), e19116.

Tseng, C. T. et al., 2010. Spatio temporal distributions of tuna species and potential habitats in the western and central Pacific Ocean derived from multi satellite data. *International Journal of Remote Sensing*, Volume 31, p. 4543–4558.

Venugopal, G., 1998. Monitoring the effects of biological control of water hyacinths using remotely sensed data: A case study of Bangalore, India. *Singapore Journal of Tropical Geography*, Volume 19, p. 92–105.

Vichi, M. et al., 2011. Global and regional ocean carbon update and climate change: Sensitivity to a substantial mitigation scenario. *Climate Dynamics*, Volume 37, p. 1929–1947.

Wilfong, B. N., Gorchov, D. L. & Henry, M. C., 2009. Detecting an invasive shrub in deciduous forest understories using remote sensing. *Weed Science*, Volume 57, p. 512–520.

Wilkie, D. S., Bennett, E. L., Peres, C. A. & Cunningham, A. A., 2011. The empty forest revisited. *Annals of the New York Academy of Sciences*, Volume 1223, p. 120–128.

Index

Note: Page numbers in *italics* refer to Figures; those in **bold** to Tables.

Protected Areas: Are They Safeguarding Biodiversity?, First Edition.
Edited by Lucas N. Joppa, Jonathan E. M. Baillie and John G. Robinson.